Legal Architecture

Legal Architecture addresses how the design of the courthouse and courtroom can be seen as a physical expression of our relationship with ideals of justice. It provides an alternative history of the trial, which charts the troubled history of notions of due process and participation. In contrast to visions of judicial space as neutral, Linda Mulcahy argues that understanding the factors that determine the architecture of the courthouse and courtroom are crucial to a broader and more nuanced understanding of the trial. The partitioning of the courtroom into zones and the restriction of movement within it are the result of turf wars about who can legitimately participate in the legal arena and call the judiciary to account. The gradual containment of the public, the increasing amount of space allocated to advocates, and the creation of dedicated space for journalists and the jury, all have complex histories that deserve more attention than they have been given. But these issues are not only of historical significance. Across jurisdictions, questions are now being asked about the internal configurations of the courthouse and courtroom, and whether standard designs meet the needs of modern participatory democracies. The presence and design of the modern dock; the dematerialisation of the courtroom by increasing use of new technologies; and the extent to which courthouses can be described as public spaces are all being hotly debated. This fascinating and original reflection on legal architecture will be of interest to socio-legal or critical scholars working in the field of legal geography, legal history, criminology, legal systems, legal method, evidence, human rights and architecture.

Linda Mulcahy is a Professor of Law at the London School of Economics.

Legal Architecture

Justice, due process and the place of law

Linda Mulcahy

a GlassHouse book

First published 2011
by Routledge
2 Park Square, Milton Park, Abingdon, Oxfordshire OX14 4RN
Simultaneously published in the USA and Canada
by Routledge
711 Third Avenue, New York, NY 10017
A GlassHouse book

Routledge is an imprint of the Taylor & Francis Group, an informa business

First issued in paperback 2011

© 2011 Linda Mulcahy

The right of Linda Mulcahy to be identified as author of this work has
been asserted by her in accordance with sections 77 and 78 of the
Copyright, Designs and Patents Act 1988.

Typeset in Times New Roman by
RefineCatch Limited, Bungay, Suffolk

All rights reserved. No part of this book may be reprinted or
reproduced or utilised in any form or by any electronic,
mechanical, or other means, now known or hereafter
invented, including photocopying and recording, or in any
information storage or retrieval system, without permission in
writing from the publishers.

British Library Cataloguing in Publication Data
A catalogue record for this book is available
from the British Library

Library of Congress Cataloging in Publication Data
Mulcahy, Linda, 1962–
  Legal architecture : justice, due process and the place of law / Linda
Mulcahy.
    p. cm.
  "A GlassHouse book."
  Includes index.
  ISBN 978–0–415–57539–3
  1. Justice, Administration of—England. 2. Courthouses—England. I.
Title.
  KD7100.M85 2011
  347.42—dc22
                                                        2010022754

ISBN 13: 978–0–415–57539–3 (hbk)
ISBN 13: 978–0–203–83624–8 (ebk)
ISBN 13: 978-0-415-61869-4 (pbk)

For Phoebe and Seamus

Contents

List of figures and sources	x
Cases	xi
Legislation	xiii
Acknowledgements	xiv

1   Architects of justice   1

Introduction 1
Existing work in the field 3
Methodology 5
Key themes in the book 6

2   An ideal type? Visions of the courthouse over time   14

Introduction 14
Justice without walls 15
Ceremony 17
Containment 21
Thresholds 22
Shared public spaces 24
Ritual and place 27
Parallel functions 29
A distinct domain 32
Conclusion 33

3   Segmentation and segregation   38

Introduction 38
The vulgar and the sacred 39
Segmentation 43
Segregated circulation routes 48

viii   Contents

What motivates segmentation and segregation of participants? 51
Conclusion 56

4   Presumed innocent?   59

Introduction 59
Differentiation 60
The spatial demands of lawyers 64
Isolation 67
Challenges to the isolation of the defendant 73
Degradation 74
Conclusion 78

5   Open justice, the dirty public and the press   83

Introduction 83
The principle of open court 84
Keeping the court open 87
Modern-day practices 95
Caught reporting 97
Putting the press in their place 101
An ambivalent relationship 103
The reinvention of public space 104
Conclusion 106

6   The heyday of court design?   112

Introduction 112
A taste for comfort 113
Temples to justice 118
Symbolic courts and civic pride 124
The role of law in emerging cities 126
Fear 131
Conclusion 134

7   Back to the future: Is there such a thing as a just court?   139

Introduction 139
Centralisation of design 140
The current state of building stock 145
New visions of a democratic future? 151
Conclusion 159

8 The dematerialisation of the courthouse 162

Introduction 162
Technology in the court 164
Law's aesthetic 169
Challenges to these assumptions 170
So what? 173
Meaningful encounters? 174
Conclusion 178

Bibliography 183
Index 201

List of figures and sources

2.1	Trees provided the setting for many trials in England until the eighteenth century (Linda Mulcahy)	16
2.2	Abingdon County Hall (1677–82) (Linda Mulcahy)	30
3.1	The fifteenth century court (Emma Rowden)	40
3.2	The consistory court in Chester Cathedral (Linda Mulcahy)	42
3.3	Segmentation at Presteigne County Hall (1826–9) (Linda Mulcahy)	44
3.4	The layout of a modern courtroom in a Crown court (Emma Rowden)	47
4.1	Nottingham Guildhall (1885–8) (Ray Teece)	65
4.2	Dock at Newcastle Guildhall (1655–8) (Steve Milor)	69
4.3	Approach to the dock from underground cells in St George's Hall Liverpool (Linda Mulcahy)	70
4.4	A modern glass dock (Her Majesty's Court Service)	76
5.1	Entrance for the public at Liverpool Sessions House (1882–4) (Linda Mulcahy)	91
6.1	Manchester Assize Courts (1859–65) (Manchester Archives and Local Studies)	118
6.2	St George's Hall in Liverpool (1841–7) (Linda Mulcahy)	120
6.3	Leeds Town Hall (1853–8) (Leeds City Council)	122
7.1	Modern courts are now much flatter than their Victorian counterparts (Her Majesty's Court Service)	143
7.2	Foyer of the South African Constitutional Court (Ben Law-Viljoen, David Krut Publishing)	156
7.3	Seating in the public waiting area outside the courts at the newly constructed Manchester Civil Justice Centre (Linda Mulcahy)	157
7.4	The Commonwealth Law Courts, Melbourne (John Gollings)	158
8.1	Video suite in Manchester Civil Justice Centre (Linda Mulcahy)	171
8.2	McGlothlin Courtroom at the Center for Legal and Court Technology (McGlothlin Courtroom, Center for Legal and Court Technology)	176

Cases

Black v Pastouna [2005] EWCA Civ 1389	166, 179
Bremer Vulcan v South India Shipping Corporation [1980] 1 All ER 439	80
Bremer Vulcan v South India Shipping Corporation [1981] AC 909	80
Bumper v Gunter, 635 F.2d 907 (1st Cir.1980)	82
Chandler v Florida, 449 U.S. 560, 101 S. Ct. 802, 66 L. Ed. 2d 740	111
Commonwealth v Moore, 397 Mass. 106, 111, 393 N.E.2d 904 (1979)	82
Delta v France (1993) 16 EHRR 574	180
Estes v Texas, 381 U.S. 532, 85 S. Ct. 1628, 14 L. Ed. 2d 543	111
Henderson v SBS Realisations, 13 April 1992 unreported. See Official Transcripts 1990–92 on www.lexisnexis.co.uk	180
King v Wilson 3AD and E 827	180
Kostovski v Netherlands (1990) 12 EHRR 434	180
Musladin v Lamarque (2005), United States Court of Appeals for the Ninth Circuit, No 03-16653, D.C. No. CV-00-01998-JL	107
People v Zamora (2000) 230 Cal.App.3d 1627	82
Polanski v Conde Nast Publications Ltd CA [2003] EWCA Civ 1573	166, 172, 180
Polanski v Conde Nast Publications Ltd HL [2005] UKHL 10	180
Queen v Abdul Nacer Bebrika and others (Ruling no 12) [2007] VSC 524	76
R (on the application of D) v Camberwell Green Youth Court [2005] UKHL 4	179
R v Carlile (1831) 6 Car and P 635, p1397; 172 Eng. Rep. 763	73, 74
R v Douglas (1841) 1 Car and M 193; 174 Eng. Rep. 468	82
R v George (1840) 9 Car and P 484, 173 Eng Rep. 922	73, 74, 82
R v Lovett (1839) 9 Car and P 452; 17 Eng. Rep. 315	73
R v Tooke, 25 State Tr. 1 at p.6	82
R v Vincent, Edwards, Drinkwater and Townsend (1839) 9 Carrington and Payne 91 173 ER 754	74
R v Vipont and others (1761), 2 Burr 1163, p. 767; 97 Eng. Rep. 767	180
R v Zelueta (1843) 1 Car and K 215; 174 ER 781	73, 82
Staley v Hunt (1834) as reported in *Rex v Carlile* (1834) 6 Carrington and Payne 636 172 ER 1397	74

Smellie (1919) 14 Cr App R 128	179
Unterpertinger v Austria (1991) 13 EHRR 175	180
V v United Kingdom and *T v United Kingdom* (2000) 30 EHRR 121	75
Wakefield's Case 1 Lewin 276	82
Walker v Butterworth, 599 F.2d 1074, 1080 (1st Cir.1979)	77
Windisch v Austria (1991) 13 EHRR 175	180
Young v Callahan (1983) 700 F.2d 32	77

Legislation

Act of Settlement 1701
The Children Act 1908
Children and Young Persons Act 1933
Common Law Procedure and Chancery Amendment Act 1852
Courts Act 1927
Criminal Justice Act 1925
Criminal Justice Act 1988
Criminal Justice Act 2003
Gaol Act 1823
Judicial Proceedings (Regulation of Reports) Act 1926
Libel Act 1792
Libel Act 1843
Matrimonial Causes Act 1857
Newspaper Stamp Duties Act 1819
Official Secrets Act 1989
Penitentiary Act 1799
The Perjury Statute 1563
Police and Justice Act 2006
Printing Act 1695
Prisoner's Counsel Act 1836
Reform Act 1832
Security Service Act 1989
Supreme Court of Judicature Acts 1873–5
Treason Trials Act 1696
Youth Justice and Criminal Evidence Act 1999

Acknowledgements

My thanks go to a number of friends who have played a part in bringing this project to fruition. The book was conceived while I was at Birkbeck and the many friends I made while there provided an ideal research environment in which I could flesh out my ideas. Les Moran and Costas Douzinas were particularly generous with their time and have acted as sounding boards for many of the ideas presented in the pages that follow. Particular thanks also go to the judges and architects with whom I have visited courthouses in England, France, Belgium and Australia over the last five years under the auspices of the Court of the Future Network. My thanks to all of you for the insights I have gained by listening to you. David Tait and Emma Rowden have provided encouragement, support and friendship throughout the project and been generous in their introductions to others who share my interests. Judith Resnik has provided me with many a valuable insight and has been generous in sharing the proofs of her book which covers some of the same ground as my own from an American perspective. Jonathan Fuller and Josie Lloyd have provided valuable research assistance.

I would also like to express my gratitude to Stephen Quinlan and Denton Corker Marshall Architects; Mr Nick Fry, Heritage Manager, Chester Cathedral; Ray Teece; Manchester Archives and Local Studies; Leeds City Council; Ben Law-Viljoen, David Krut Publishing and the Constitutional Court of South Africa; John Gollings and Hassell Architects; Sara Edet and Lesley Armstrong of Her Majesty's Courts Service; Staff at the museum in Abingdon County Hall and Fred Lederer of the Center for Legal and Court Technology for allowing me to reproduce images over which they hold the copyright or take photographs in buildings which they manage. Recognition should also extend to the many court clerks and ushers who have provided me with invaluable information about the day to day workings of the courts and given of their time to show me round some of them.

The production of this book has taken me on a methodological journey during which I have striven to produce a more nuanced account of buildings than I was previously capable of. As I have written I have also become a student of architectural history. For an academic lawyer trained in interpreting texts, learning to 'read' a building and interrogate the image has been liberating and I am grateful to my tutors for their patience in dealing with my obsession with court architecture. Finally, I offer my gratitude to my husband Richard who has cooked and cleaned for me as I completed this book. My children, Connor and Sam, will no doubt have left home before I complete my next monograph but I am afraid that Richard will live to enjoy the experience again.

Chapter 1

Architects of justice

Introduction

This book presents an 'alternative' history of the trial. In contrast to previous accounts of legal proceedings which rely on ethnographic studies, an analysis of reported cases or consideration of the jurisprudence of procedure and outcome this monograph tells its story through the architecture of the courthouse. It will be argued that the environment in which the trial takes place can be seen as a physical expression of our relationship with the ideals of justice but that despite its importance the geopolitics of the trial has received very little attention from academics. In contrast to a vision of judicial space as neutral this book argues that understanding the factors which determine the internal design of the courtroom are crucial to a broader and more nuanced understanding of state-sanctioned adjudication. The containment of the jury, the increasing amount of space allocated to advocates, the incarceration of the criminal defendant in the dock, the containment of spectators and the creation of dedicated space for journalists all have complex histories which deserve to be charted and discussed much more than has been the case to date. Each time a section of floor is raised, a barrier installed or a segregated circulation route added it has the potential to create insiders and outsiders; empowered and disempowered participants in a space ostensibly labelled 'public' in which the intricacies of civil liberties and participatory democracy are played out.

We readily think of trials as ritualised events performed according to a social and legal script conferring authority and resulting in documented outcomes. Official language, prescribed procedure, codes of conduct and documentation are recognised as conferring legitimacy on such proceedings but the legitimacy of the trial also derives from the setting in which these rituals take place. Public buildings can both inspire and degrade those within them; they can calm or oppress. The spatial configurations of the courthouse and courtroom can confer prestige or dignity on those who use them or serve to undermine their credibility. Legal architecture can associate law with tradition and conservatism or can equally well symbolise a commitment to change and innovation. Courthouses can act as memorials to the past as well as reflecting aspirations for the future. In a US context, Chief Justice Hennessey (1984) has argued:

> Our courthouses are monuments to our legal tradition, its noble purposes and occasional tragic miscarriages. They evoke the memory of historical events and of the aspirations, frustrations and fears of the many people – the learned, the dedicated, the articulate, the oppressed and the despised, the avaricious and the brutal – whom the law has summoned to exercise their skills or to account for their actions . . . they are not merely buildings, rooms and furniture but are, rather, monuments that evoke several centuries of human effort and progress.

Architects and those commissioning buildings have long understood the importance of space and place in creating and reinforcing courtroom identities but this study of court architecture encourages the reader to confront the interface between rhetoric and reality. Histories of civil liberties, punishment and procedure have tended to focus on the ways in which the trial has 'evolved' as the defendant, the press and the legal profession have acquired rights to speak, report and defend. From a position in which a case might be decided by chance or according to the whim or prejudice of an adjudicator we tend to celebrate the modern trial as a rational process in which the rights of individuals are better protected and excesses of partiality impossible. My aim is not to challenge the thrust of such accounts but an analysis of the spatial configurations of the courthouse and courtroom makes clear that the segmentation and segregation of space has often served to undermine civil liberties and restrict effective participation in the trial. At points the story of legal architecture has reflected evolution and revolution in criminal and civil procedure but it has also served to subvert traditional accounts of the progressive acquisition of rights by marginalising litigants and the public in proceedings as England moved towards a representative democracy.

Although this monograph draws heavily on the history of court architecture the issues it raises are far from being of only historical significance. Many of the spatial practices adopted in the courtrooms of today evolved in very different social and political contexts but are rarely subjected to sustained critique. Legal systems throughout the world draw heavily on traditional practices as a way of conferring gravitas and cultural meaning on proceedings but there is a danger that an over reliance on historical precedents can transform justice facilities into frozen sites of nostalgia (Graham 2004). Important challenges about the contemporary relevance of the placing of participants in the trial await those who commission and design courthouses. Are modern courthouses in which most of the space is not accessible to the public appropriately labelled public buildings? Does the appearance and positioning of the dock undermine the presumption of innocence in the English trial? Is it appropriate for legal counsel in superior courts to sit with their back to their client? Is the increasing use of video link in danger of rendering the modern trial an inauthentic legal ritual? If the state of current building stock reflects the respect in which the legal system is held by the State can we assume that its significance is

Architects of justice 3

in decline? In the chapters which follow I attempt to address these and many related questions.

Existing work in the field

The social significance of court architecture has long been neglected by academics. In the field of legal scholarship, the absence of research on the use and experience of courthouses and courtrooms can, in part, be explained by lawyers' obsession with the word. When we teach our students about law we do so through the medium of the written judgment or transcript as though they give a complete account of why a case is decided in a particular way. We frequently assume that if all else is equal that judgment given in one place would be the same as judgment reached in another. This conceptualisation of the legal arena undoubtedly limits our appreciation of how spatial dynamics can influence what evidence is forthcoming, the basis on which judgments are made and the confidence that the public have in the process of adjudication. It is also the case that to date, lawyers' understanding of the origins of the modern trial and the history of ideas about it stem largely from accounts of atypical criminal trials such as records of *The State Trials*. Excellent use has been made of the Old Bailey sessions papers but again, these present a rather London-centric account of the English trial.[1]

Studies of courthouses have received slightly more generous attention from other disciplines but the literature remains limited.[2] Architectural historians have tended to lavish attention on other public buildings such as churches, castles, prisons or town halls to the detriment of discrete studies of the courthouse. Technical accounts of historic courthouses which focus on aesthetic convention or style such as those provided by English Heritage, the Pevsner guides and the Victorian County Histories are informative but tell us very little about the social or political significance of the spatial practices described. For some commentators the ban on photography of any kind inside law courts has contributed to the under-appreciation of the architecture of law courts (SAVE 2003).[3] It has also been the case that traditionally architectural historians have tended to focus on well known architects rather than particular building types. The result is that discussions of courts only occur when they have been designed by the renowned.[4] Contemporary architectural historians have been more prepared to go behind appreciation of technique and style to an understanding of what buildings symbolise, their setting, how they came to be and why they are the way they are. Mark Girouard's (1990) study of the English town is an excellent example of this genre which provides numerous insights into the problem of housing the twice-yearly Assize courts.[5] But other accounts of legal architecture have tended to focus on particular symbolic courts of national significance. Research monographs in this category include Brownlee's (1984) excellent account of the Royal Courts of Justice, Sharon's (1993) book on the Israeli Supreme Court, Burklin *et al*'s (2004) account of the design of the federal Constitutional Court of Germany and Pevsner's (1976) short review of notable courts.

4 Legal Architecture: Justice, due process and the place of law

Discussion of trends in everyday courthouse design are in much shorter supply. Chalklin's (1998) detailed account of English Counties and Public buildings provides an invaluable index of the courts constructed from 1650–1830 and the cost of their construction and McNamara (2004) charts lawyers' and architects' involvement in the development of the American court from tavern to courthouse over a similar period. Two academic monographs are worthy of particular mention for placing court design in a broader legal and political context and have been of considerable influence in my own research. Katherine Fischer Taylor's (1993) book on the architectural, historical and social significance of the Parisian Palais de Justice has broken new ground in the field with its sophisticated account of the link between architecture and political theory in the post-Revolutionary era in France. Her work encourages us to look at courthouses as cultural icons which can be used to chart important shifts in ideology. Clare Graham's (2003) book on the architectural and social history of English law courts to 1914 is the first comprehensive attempt to understand the social significance of court design over time and is invaluable to any scholar attempting to plot significant trends. The breadth of her book is impressive, charting as it does various adjudicatory settings such as coroner's courts, petty sessions, quarter sessions, police, county and Assize courts which have needed housing over time. The value of the gazetteer of court buildings which forms the final section of the book can probably only be truly appreciated by those of us who have ventured into County records offices in search of research leads.

One might have expected sociologists and social geographers to have been more interested in the geopolitics of the trial as a prime site of state control of the individual but while control of territory has long been seen as fundamental to studies of power dynamics in society it is only recently that these disciplines have turned their attention to the interface between law, place and space (see Foucault 1977; Blomley *et al* 2001; Harvey 1996; Evans 1999; Fairweather and McConville 2000). Whilst these studies begin to provide us with the conceptual tools to understand the interfaces between legal architecture and power very little attention has been given to the law court (but see Dovey and Fitzgerald 2010). One might expect the use of space within the courtroom to be a central concern of criminologists but in recent decades their attention has tended to focus on the range of sites outside of the courts in which criminal justice is administered and punishment meted out.[6] Two enthnographic accounts of space and place in the courtroom stand out as exceptional in this context. Pat Carlen's (1974, 1976) study of the spatial dynamics of Magistrates' proceedings, which she characterises as a theatre of the absurd, encourages us to look at the ways in which spatial configurations in the courtroom impact upon and degrade lay participants. Paul Rock's (1991, 1993) seminal study of Wood Green Court suggests that the courthouse is best understood as a symbol of social organisation which manages and confirms standardised identities.

This book seeks to steer a course between these various approaches in order to unravel the connections between architectural style, political and social change, concepts of the public sphere and histories of the trial. Written from a socio-legal perspective my intention is not to produce a history of the theory of criminal and civil procedure nor a sophisticated account of the law of evidence or civil liberties. These stories intersect with my own at various stages but they are better told by those with expertise in those fields. My aim instead has been to explore what the use of space in the courthouse and courtroom tells us about the respect afforded participants and the social order of the courthouse over time. It is an account which focuses in particular on the experience of the lay participant in the trial and the various ways in which they have been, and continue to be, controlled and surveyed. In particular this book represents an attempt to plot out the complicity of architecture in classifying and containing the participants in the trial in ways which are problematic to those of us interested in the delivery of equal access to justice.

Methodology

Data about the architecture of the law court exists in a wide range of locations. Graham's (2003) gazetteer lists over 850 buildings which served as a courthouse in England up until 1914 and the number of historic courthouses which are still standing has provided me with what others have referred to as an untapped 'architectural treasure trove' (SAVE 2004, p.1). Some of these such as John Carr's Assize courts at York, Robert Smirke's court in the precincts of Lincoln Castle, Robert Elmes' magnificent St George's Hall in Liverpool and Street's gothic revival Royal Courts of Justice have been widely acclaimed as being amongst the finest buildings of their era and clearly speak to law's architectural ambitions. But for a socio-legal scholar it is the mundane as well as the magnificent which has attractions. In addition to the artistically acclaimed, this book also attempts to deal with the ordinary, the provincial, the brutal and the ugly. A major assumption of my work has been that lack of ambition reveals as much, if not more, than the monumental about attitudes to justice.

In an attempt to give depth to this study of court architecture I focus on one type of court, the Criminal Assizes.[7] My emphasis on Assize courts had a number of advantages. The first is their longevity. Judicial Assizes have been held since medieval times and offered a rich backdrop against which to examine shifts in thinking about design across centuries. For Cockburn (1972) the Assizes are the 'most familiar and longest lived of English courts' and a study of them has much to reveal about changing notions of law and procedure over time. For many centuries the Assizes were the most important courts in the provinces and the arrival of the King's representatives and itinerant judges marked the most significant event in the local political and social calendar. Until the nineteenth century the Assize courts continued to be part of regional government and conducted a mixture of legal and executive functions. As a result they become central to debates about the political compact between local and central rulers and the fulfilment of the promises of the Act of Settlement. Moreover, although the

Assizes were abolished in the 1970s Crown Magistrates and County courts continue to sit in the same buildings. Moreover, when purpose-built courthouses began to emerge as a specialist building type it was the Assize courts which tended to lead the trend for purpose-built and permanent facilities and local dignitaries were more likely to expend significant resources on the building used to house the Assize than other courts. These factors alone meant that buildings for the Assizes were likely to reveal a considerable amount about regional identity and local attitudes to justice.

In the course of my research I have visited numerous records offices and courthouses. Some of the historic courthouses I was interested in are still used for trials today and so I was able to observe the spatial dynamics of trials from the public gallery. Other courthouses such as St George's Hall in Liverpool and those at Presteigne, Bodmin, and Nottingham Shire Hall have been converted into museums and their courtrooms of the eighteenth and nineteenth century have been kept intact. My research has also involved a considerable amount of research at the Royal Institute of British Architects where I have reviewed a range of architectural journals and been granted access to plans and photographs. Extensive use has also been made of specialist legal journals at Colindale newspaper library and to other generalist newspapers published from the eighteenth century onwards. Finally I have visited courthouses in a number of other jurisdictions and have been able to talk to a number of contemporary architects of recently built courts about the ideas which have fuelled their design.

Key themes in the book

I argue that in approaching a study of the architecture of the trial it is important to view the subject as part of an ever-changing history of ideas about public and legal space. As a result, contemporary court design guides produced by Her Majesty's Courts Service which suggest that we have reached a point in which templates for courtrooms have reached a state of perfection deserve to be contested. Not only is the concept of a dedicated courthouse or courtroom a relatively recent phenomenon but certain spatial practices within the courtroom continue to be questioned across jurisdictions and to raise issues which should be questioned. New interest in restorative justice, for instance, has brought an interest in sentencing circles to replace the linear hierarchy of much contemporary design and the country's first Community Justice in Liverpool has challenged current spatial practices with its roundtable facilities and the judge's insistence that the defendant sit close by the judicial bench.

The notion of justice without walls is far from revolutionary. If one understands the study of architecture as concerning any attempt by humankind to alter their surroundings then the architecture of law courts is not synonymous with buildings. The widespread use of buildings totally dedicated to law is a relatively recent phenomenon. Circles of stones, the gates of cities, moot hills and trees have all been used as signifiers of the special places in which law has been administered since ancient times. There are records of courts in common law jurisdictions being held in a field under a tree as recently as the nineteenth century and in Australia policy makers and the judiciary are now experimenting with outdoor trials involving aboriginal members of the community once more. Historically, some outdoor locations were chosen on purely pragmatic grounds because they were easy to locate but they might also be selected because they embued the proceedings with a particular authority which drew on history, physical beauty or the belief that it was important to administer justice in the sight of God. The location of trials outside also meant that they were literally as well as metaphorically 'open' in a way which is not true of the modern trial. A trial moved indoors is one in which there is inevitably less space for those not centrally involved in proceedings and in which thresholds, entrances and barriers can be more easily used to denote status. In societies in which political power is dispersed an outdoor location which facilitated widespread involvement could provide a strong metaphor for the ideology which underpinned the legal system. Circular or open designs often symbolise a group ritual in which all participate rather than one in which the public merely spectate. Far from being of only historical significance, it is clear that the concept of walls being necessary to enclose legal proceedings is once again being challenged as new technologies make 'virtual' appearances in court from remote locations possible. Researchers have even staged trials where none of the parties to proceedings, officials or judges is present in one physical locality. The possibilities of such technologies have led some to claim that the courthouse has, once again, been dematerialised, a claim which reminds us that the notion of courthouse or courtroom is far from being a stable concept.

Historical accounts reveal that 'courthouses' and 'courtrooms' are relatively recent inventions. For many centuries trials across legal jurisdictions within England shared space with political debates, balls and assemblies, church services, markets and theatres. Even the central courts at Westminster were not accommodated in a purpose-built building devoted solely to law until the nineteenth century and today some courts continue to share facilities with other public bodies. In addition to sharing generic public buildings which lacked symbols and signs devoted solely to law it is also the case that for many centuries the civil and criminal Assize courts often shared a room when sitting. The implications of such practices are considerable. The fixtures and fittings associated with the court had to be portable, simple and suitable for other purposes. Moreover, the transient nature of courts and the fact that two Assize courts might sit at the same time in one room meant that those who wished to observe them enjoyed a freedom of movement not experienced by modern audiences. Attempts to uphold the notion of the sociable court continued long after the judiciary were allocated dedicated spaces within public buildings in which to conduct trials and until the late eighteenth century it remained common for Assize courts to be marked off from a central public hall by pillars rather than walls.

Today, when we imagine a courthouse we tend to think of a public building which uses massing, shape and style to convey a sense of importance or foreboding. Law courts are typically different from surrounding buildings and exist as culturally specific markers in the civic landscape but this is also a relatively recent development. It will become clear to the reader in the chapters that follow, that like so many other social and cultural developments in our history it is in the late eighteenth and nineteenth century that ideas about public space were revolutionised. Nowhere is this more apparent than in architecture. New standards of comfort, new types of buildings, burgeoning interest in the classical form together with new ways of thinking about the private and public domain all led to unprecedented changes in the spatial practices of English society. The law court was no exception. A new raft of *palais de justices* were constructed which separated law out from other civic functions and created dedicated and permanent spaces for trials. Not only did these serve to glorify the role of secular law in the aftermath of the enlightenment but they provided unprecedented opportunities for the placing of people in space.

The trend towards purpose-built courthouses which occurred from the late eighteenth century onwards can be described as a revolution in legal architecture which the Victorians were keen to embrace. In many ways, if judged by ambition alone, the nineteenth century represents the heyday of court design. Never before, and very rarely since, have English architects executed such ambitious monuments to law. Courthouses of this period became larger, more complex and offered a range of new facilities to those who used them. Indeed, several courthouses of this period are considered to be among the best buildings of the age. Law courts were not alone in terms of the resources lavished upon them during this era. Elegant and grand libraries, reading rooms, museums, art galleries, railway stations and hotels all became prominent in towns and cities throughout this era. In many of the emerging industrial cities of the North and Midlands civic building programmes of the time have been understood as reflecting civic pride. These accounts are compelling but it is argued here that reducing the architectural ambition of law courts to civic pride alone obscures the importance of law in the capitalist project. One narrative which is missing from these accounts of architectural ambition during the era is the particular relevance of extensive resources being expended on courts of law.

There are a number of reasons why the law was particularly worthy of being celebrated in the Victorian era. Extensive legal reforms, which spanned most of the nineteenth century led to a newly structured legal system in which much of the overlap which existed between the parallel legal systems at the dawn of the century were abolished. New layers were also added to the legal system in order to render procedures for dealing with petty crime and recovery of small debts more accessible and cheaper. Courts of this era can also be seen as symbols of the new social contract being promoted by the very industrial classes who were increasingly responsible for commissioning and overseeing many of the new courthouses. Their interests lay with the promotion of a society where new economic wealth could be achieved through the market rather than rank and it was to the law that such industrialists often turned for legitimation of free trade.

Social commentators of this time became increasingly enthusiastic about the possibilities that the newly reformed legal system would not only reflect change but promote it. Although the judiciary's support of the industrial project was far from unequivocable, it is undoubtedly the case that many of the new doctrines introduced during this period favoured the industrial classes and the concept of free market which underpinned the acquisition of much of their wealth.

While the grand monuments to law constructed from the late eighteenth century to the end of the nineteenth gave a certain dignity and gravitas to legal proceedings the impact on those standing in the dock or witness stand was likely to be very different. What might appear inspiring to stakeholders in the newly reformed legal system or industrial society could just as easily be seen as forming part of degradation ritual for those subjected to the violence of law. As well as being the age of reform the nineteenth century was also a period in which there were constant fears about working class unrest fuelled by working and living conditions and the many recessions which occurred. Viewed from this perspective the new law courts could just as easily be seen as crude assertions of authority over the threatening masses. Indeed, one of the most important narratives to emerge from this book is the extent to which fear has motivated changing concepts of court design. Despite the emphasis on progressive narratives elsewhere, the containment of certain participants in the trial has actually become more pronounced as time has passed. The architecture of the courtroom has transformed from what can be characterised as a 'social court' (Graham 2004) to a highly segmented and segregated space in the first decade of the twenty-first century.

These shifts in court design must be understood in the context of changing notions of due process. Changes to the law of evidence and emerging codes of professional etiquette have been motivated in part by concerns about the contamination of testimony and intimidation of participants in the trial. Viewed in this way, dividing the courthouse and courtroom into segments and creating segregated circulation routes can be seen as a way of increasing the legitimacy of the trial by protecting the vulnerable from unsuitable encounters. But while segmentation and segregation have meant that some participants in the trial are offered a retreat from the public arena or protection from shared space, for others it has led to their increasing containment within designated zones in the courthouse and characterisation as dangerous. The position is such that it is no longer accurate to refer to the courthouse with its numerous private zones as a public building. This thesis becomes particularly potent when one turns to look at the placing of certain participants in the criminal trial such as the defendant and their counsel. From a position in the fifteenth century when counsel and prisoner stood shoulder to shoulder at the bar of the court when presenting their case, lawyers now sit before their clients separated from them by physical barriers and with their back turned on them. Not only has the defendant become increasingly isolated in the courtroom but in the great monuments to law of the nineteenth century barristers also began to enjoy special spatial privileges in the environs of the court. As they were provided with private dining rooms, robing rooms, libraries, rooms for consultations and dedicated spaces for their clerks, defendants became slowly incarcerated in fortified docks linked to cells by dismal underground passages. These practices continue today with the result that the modern dock could be viewed as a brand of incarceration which militates against the presumption of innocence. Despite this, and in contrast to the legal systems in countries such as America and Sweden, the Ministry of Justice in the UK remains committed to docks placed at the margins of our courtrooms in Crown courts. Indeed some now take the form of glass rooms within the courtroom which serve to reinforce the isolation of the defendant and undermine their ability to engage in their trial.

The role of the public in the trial is also ripe for debate. Most developed societies assert that the public play an important role in legitimising legal proceedings. For many, open justice is treated as synonymous with the notion of a fair and accurate trial because it provides important checks on the credibility of witness testimony and the partiality of the judge. Others have justified the admission of the public to the trial on the basis that it educates spectators in the ways of the law. Despite such aspirations, spatial practices in the courthouse would also appear to have been fuelled by distrust of the public. Fear of uncharted or unscripted performances in the environs of the court emerge as dominant narratives in discussions of court design. The implications of this are considerable. In the chapters which follow I suggest that the notion of an open court can only be meaningful where the public have a keen sense that they are participants in the public ritual of the trial and that they have an important role to play in calling the legal system to account. But in reality, the principle of open justice is in serious danger of being undermined when the public are first searched, then channelled through space and finally positioned within tightly controlled zones within the courtroom. While it may be the case that much more dignity is accorded participants in the trial over time we have not yet reached a state in which the courthouse can be described as genuinely open. Historical accounts of the trial and its architecture remind us that since the eighteenth century the court has moved from a position in which it was housed in multi-purpose public space frequented by the public on a regular basis to one in which the trial is set apart in specially constructed buildings designed by reference to guidelines which treat the public as visitors to be wary of. In the modern courts, surveillance techniques, the positioning of screens on which evidence is shown, the interruption of sightlines between the public and key participants and the containment of the public within glass rooms within the courtroom all speak to the ongoing marginalisation of the public and their transformation from active participants to docile bodies.

The position of the public in the trial has been intimately connected with the fate of the press. The latter have fought hard for the right to attend, comment on and report proceedings and it is undoubtedly the case that they have played a significant role in protecting the public interest by uncovering miscarriages of justice and procedural error. Their place in the modern trial is reflected in contemporary design guides which recognise the legitimacy of the press demanding dedicated facilities in the courthouse. Newspaper and television journalism has also played a part in disrupting traditional notions of the courthouse and courtroom by bringing the drama of proceedings into the living rooms, television sets and laptops of the public where their consumption of legal narratives can not be monitored by a judge or court staff. However, these achievements have prompted new tensions between the interests of the press and the public interest. In many ways the participation of the press in the trial has led to the further marginalisation of the public as their presence became a substitute for a genuinely open court. Somewhat ironically given the part the press have played in encouraging interest in trials, members of the public have also at times attracted criticism from the press as a result of their macabre enthusiasm for news of sensational trials.

Arguments about the extent to which the interests of the public, the press and the legal system can be treated as synonymous continue to be discussed in the context of the televising of trials and who should control the images produced. The media circus which can follow the filming of sensational trials can appear to degrade the legal system and judicial conservatism about filming suggests that adjudicators like to know who is 'in' the court. By the same token, the failure on the part of the UK government to make a policy decision about televising trials is difficult to justify in an era in which politicians and the court service are embracing the use of live link to transmit the evidence of defendants and witnesses into the court from afar. It is increasingly difficult to understand why the use of technologies to bring testimony into the court is acceptable while its use to transmit images of the trial to the laptop or living room of the populace is not. These issues continue to raise difficult questions about who is benefiting from new technologies and whether they are being used to advance the public interest.

The increasing use of technology in the courtroom clearly has the potential to disrupt traditional notions of the trial. As the physical boundaries of the courtroom are increasingly violated by new information highways, ideas of 'legal place' are becoming unsettled and the legal rituals, traditionally seen as critical to the trial, marginalised. 'Wired' courtrooms now allow senior counsel in Australia to receive transcripts of ongoing proceedings in the UK within minutes and to dictate the questions being posed by a junior in the live trial. 'Smartboards' allow for computer-generated reconstructions of events to be produced which could previously only be imagined. Real-time transcription alters the pace and experience of the trial as advocates no longer pause for judicial note-taking. The orality of the trial is further compromised as judges are presented with a transcript of testimony seconds after it is delivered in person before them. While some of these innovations are still only dreamed of in the UK, all of them have been introduced within common law jurisdictions.

These changes call for more extensive public debate than has so far taken place about the architecture of the modern courthouse. There are certainly signs that courtroom drama in which the parties gather together in each other's physical presence could become a thing of the past and that our prime site of adversarial legal practice is in danger of being dematerialised. A key question posed in this book is whether the sort of trials now being imagined render the trial an 'inauthentic' legal ritual? In the new digitally mediated environments being envisaged, and slowly realised by policy makers and 'techno-evangelicals', the character and content of legal discourse is being transformed. Encounters within the courtroom are in danger of becoming sanitised as participation in the trial becomes akin to a fleeting televisual encounter. Conceptions of the court as centrally located, locally anchored, spatially discrete and architecturally symbolic are set to change. The result is a radical reorganisation of the challenges that 'spatiality' poses to law. Technology undoubtedly has the potential to enhance democratic participation by promoting discussion amongst the marginalised and geographically remote. Participation in a trial can be degrading and inhibiting, cross-examination brutal and the aftermath of proceedings traumatic. However, what is argued here is that beneath the inadequacies of adversarial adjudication there may continue to be some ideals enshrined in court practices which we want to retain in a technological age. In an era in which reform of the litigation system has been fuelled by considerations of efficiency and proportionality, it is argued that we are in danger of forgetting that requiring the physical presence of people in the court continues to have cultural resonance.

A book of this kind would be incomplete without making some suggestions about how the notion of a 'just' or democratic court might be realised. As a result I also turn to look at how the challenges in court design mooted above are being addressed in England and other jurisdictions. In many ways I end where I begin by revisiting the importance of ritual and place in the face of technological advances. Drawing on visits to courts in five legal jurisdictions I present a number of examples of modern courts which attempt to disrupt the canon of traditional design in favour of creating spaces which reflect contemporary debates about participatory justice. I seek to address how we organise the space of the courthouse and courtroom so that we can confer equal dignity on all participants, the ways in which design can be complicit in producing temples of justice which reflect the ongoing importance of gathering places in a society which aspires to an active public sphere and the part that architecture can play in undermining the alienating tendencies of the modern trial. Across jurisdictions important questions are beginning to be asked about the internal configurations of the courthouse and courtroom and what we want it to be. The issue of how we recognise a court and what architecture can tell us about the role of law in a given society at a particular time raises important concerns about how law is conceived of and its temples recognised across time and cultures. These issues are becoming critical as justice facilities are increasingly designed by international architects who come from jurisdictions where different normative values underpin the legal system. It is also the case that the growth in transient populations means that courthouses will be experienced by those who may approach the legal with different concepts of justice, due process and the State. In closing, I argue that contrary to the suggestion made in contemporary design guides the notion of what constitutes a courthouse is very far from being fixed in the imaginations of those who design them.

As the notion of public space and the public sphere become increasingly ambivalent there are undoubtedly important debates to be had about the transformation of the courtroom and its liberation from historical practices which undermine the notion of a healthy civic sphere.

Notes

1 Other historical sources such as the Assize rolls held in county records offices across the country vary considerably in quality and detail and can take considerable time to transcribe and contextualise. As a result they have been avoided by many scholars.
2 So, for instance, in his otherwise influential work account of architecture in Britain 1530–1830, Summerson (1986) pays no attention to the law court and the same is true of his account of the new building types which emerged in Britain in the eighteenth century in which he prefers to focus instead on theatres, libraries, museums, hospitals and prisons at the expense of a number of major commissions for courts.
3 See further The Courts Act 1927.
4 An example of this approach is the otherwise excellent book on Alfred Waterhouse produced by Cunningham and Waterhouse (1992) which to my mind is the only source which gives a detailed description of the building of Manchester Assize Courts and indicates Waterhouse's involvement in several other legal architecture projects.
5 Girouard (1990) paints a sophisticated account of how legal space shares its history with communal public spaces but unfortunately he leaves his account at the very point at which the architecture of courts enters into the modern age.
6 A review of work which has appeared in the *British Journal of Criminology* suggests that there is some recent scholarly work on youth courts but remarkably little research on other trials.
7 The extent of the material available has led to inevitable limitations being imposed on the scope of this book. It would require much greater resources than those available to me to visit, analyse and research the hundreds of court buildings which still stand. Whilst some plans and accompanying materials are available in the National Archives and other national collections such as those held at the Victoria and Albert museum, it is also the case that much of the documentation relating to courts prior to the 1970s is contained in the 90-odd records and archive offices across England.

Chapter 2

An ideal type? Visions of the courthouse over time

Introduction

At a 2009 conference the architect Stephen Quinlan delivered a plenary lecture on the recently built and much discussed Civil Justice centre in Manchester. This new court complex is the largest to have been built in England since the Royal Courts of Justice were completed in 1882. The postmodern glass-fronted Manchester Court complex with its nine-storey atrium and gravity-defying protruding 'fingers' could not provide a more stark contrast to George Edmund Street's ornate gothic revival building in the Strand. But what was most significant about Quinlan's presentation for present purposes was that the first comment in response to it came from an architectural historian in the audience who was concerned that the new Manchester centre was not recognisable as a courthouse. Devoid of the form, signs and symbols which we associate with this established building type, he claimed that England's most important new court building served to render the notion of 'courthouse' meaningless.

The idea that courts are, and should be, instantly recognisable is not a controversial one. Whilst recognising the variety of forms adopted by designers, Graham (2004) surmises in her authoritative account of the history of the English law court until 1914 that one of the key characteristics of modern courthouses is that their planning is 'highly distinctive' (p.1), indeed 'fossilized' (p.3). In their recent review of historic English courts for English Heritage Brodie *et al* (unpublished) have also claimed that law courts are designed to be recognised externally and understood internally. The most recent edition of the *Court Standards and Design Guide* (Her Majesty's Courts Service 2010) is equally confident about the existence of a template when prescribing design models:

> [C]ourtroom layouts . . . are the result of careful consideration by numerous user groups. They incorporate specific and well-defined relationships between the various participants by means of carefully arranged sight-lines, distances and levels. It has been found that attempts by individual designers to improve on these layouts have rarely been successful and consequently these layouts (including accompanying levels) are to be adopted in all cases. (p. 7: 1.2)

In this chapter, I argue that these assertions suggest a certainty about the settled nature of design which is easily disrupted when subjected to closer scrutiny. While the *Court Design Guide* suggests that our ideas about what constitute a court are now quite fixed my analysis suggests that the concept of a courthouse to which they allude is actually a very recent invention.

In-depth research suggests that the history of spaces dedicated to community or state-sanctioned forms of adjudication is a long one in which a wide array of spatial configurations have been designated suitable for the administration of justice. One approach to such diversity has been to arrange courts into historical clusters, each of which constitutes a 'development' on the last. Girouard's (1990) four-part chronology of the development of the modern courthouse is an example of such an approach and Graham's (2004) monograph largely presents a more detailed account of his typology. What the former lacks and the latter can often only hint at given its scope is the history of ideas which underpin shifts in design. I argue that these otherwise informative accounts tend to treat certain ideas about courthouse design as being synonymous with particular periods. While this approach has many merits, it could be suggested that what is most interesting about histories of courthouse design is the ebb and flow of ideas about legal space.

Justice without walls

In the West concepts of the trial tend to be treated as symbiotic with the enclosed places in which they take place, be it the court *house* or court *room*. But adjudication is far from dependent on a building. It is interesting to note for instance that despite their love of litigation ancient Athenians had no buildings dedicated solely to law. Indeed if architecture is defined as any attempt by humans to mould their physical environment then our understanding of the place of the trial must go beyond buildings. In his account of an ideal adjudicatory setting in the *Iliad*[1] Homer describes the seats of polished stone in the sacred circle of stones where the elders decided disputes in the open and in his research on ancient courts in Western Europe Bellott (1922) makes clear that all Teutonic and Celtic courts were originally held outside. In the British Isles numerous stone circles appear to have been used by ancient tribal communities as places in which assemblies, worship and adjudication would take place. So, for instance, it has been argued that Stonehenge was known in the Cornish dialect as Merddin Embys, which signified the fence of judgment. Indeed Bellott (1922) argues that the flat square stones which are often found in the ancient stone circles which exist across the British Isles are often referred to as an altar but were probably used as frequently just as a seat of judgment.

Moot hills, easily identifiable from afar, have also been popular settings for the resolution of disputes and other assemblies in England. Indeed, manmade moot hills are possibly amongst the first known attempts by man to shape the environment into a suitable setting for important gatherings.[2] In describing the courts of the twelfth century Maitland (1911) makes clear that the association between the

16 Legal Architecture: Justice, due process and the place of law

Figure 2.1 Trees provided the setting for many trials in England until the eighteenth century © Linda Mulcahy.

administration of justice and the moot hill has a much longer history than that of the court building or room:

> ... it may be a county court or a hundred-court, or a court held by some great baron for his tenants. It is held in the open air – perhaps upon some ancient moot-hill, which ever since the time of heathenry has been the scene of justice (p. 122)

The dispensing of justice under trees is another example of an outdoor location commonly used for adjudication. De Joinville describes Louis IX of France's practice of holding court under a tree in the thirteenth century (Graham 2004) and the use of trees as a setting for the trial is far from being a historical nicety. The Victorian County Histories record numerous examples of ash trees providing the setting for manorial courts[3] and record that the court for the Hundred was held in Bruton field in Somerset under an ash tree by three crossways as late as 1735.[4] In an American context it has been suggested that courts were held under trees for want of appropriate accommodation in some locations in 1841 (Edwards 1993). The adoption of the motif of justice under a tree by the Constitutional Court of South Africa (1997–2003) makes clear that the connection between justice and trees continues to have considerable cultural resonance in some developed societies. In the case of South Africa justice under a tree is more than just an emblem for the court. The brief given to the architects was to create a building rooted in the African landscape. In line with this, the foyer of the court is a spacious light-filled area punctuated by sloping columns which provide an architectural metaphor for the tree. In addition the roof has slots designed to create the idea of dappled sunlight filtering through leaves.[5] Closer to home, it has been argued that the wooden panelling which continues to adorn most of our courts in England, and the still popular wooden canopies over the judicial dais, can trace their origin to the times when English monarchs dispensed justice under the boughs of a tree (Jacob 1995–6).

Ceremony

It would be wrong to assume that a lack of physical props rendered procedures informal or less directed to the solemnity of the trial. Work on the aesthetics of ancient assembly sites is limited, but Watson (2001) reminds us that open air locations for significant gatherings were often carefully chosen to create a sense of spectacle or to reflect the social order of those using the site. He argues that just as modern buildings affect the behaviour which takes place within them so too would the arrangement of outdoor sites have influenced the movement of people. One goal in selecting sites would commonly have been to remove participants from the familiar and to construct a distinctive or special place for adjudication. Outdoor locations often have symbolic importance or even magical qualities (Rogge 1968–9). Graham suggests that certain locations such as Pennenden heath near Maidstone which have hosted trials for hundreds of years acquire a particular mystique. She describes them as 'a last enfeebled remnant of the ancient shire assembly, suggesting antiquarian respect for a venue hallowed by long use, perhaps even a feeling that it has accumulated its own numerous powers' (p.41).[6] In a more recent example, she describes how the formalities and rituals of trials in the higher courts of the nineteenth century built up to such a degree that the 'trial acquired an almost dream like quality, suspended outside real time' (p.271). Places also adopt the quality of being hallowed in other ways. Such was the importance of the commitment to a set place for courts demanded by the Magna Carta that one Lord Chief Justice of Common Pleas in the sixteenth century would not move court furniture one foot to keep the participants out of the draught for fear of invalidating the proceedings (Maitland 1883–84).

The organisation of an outdoor setting could also be used to symbolise the political ideological underpinning proceedings. Courts arranged in a circle, such as the one described by Homer were likely to be considered more appropriate in tribal societies in which formal power was dispersed. Circles were commonly used to symbolise the fact that justice was administered in the name of the whole community.[7] The presence of inner and outer circles tended to denote that in tribal communities groups of wise men were acknowledged as guardians of the customary traditions transmitted orally from one generation to another. As law makers, they identified the appropriate rule and proposed judgment to the popular assembly but it was the assembly as a whole that gave judgment and took responsibility for it (Stein 1984). The ongoing symbolic importance of circular arrangements for proceedings continues to be emphasised in courthouses today and in alternative dispute resolution forums such as mediation. The sentencing circles employed in some courts in Australia and Canada in which those intimate with the accused participate in discussions about what constitutes an appropriate punishment on conviction are just one such contemporary example which the Community Justice Centre in Liverpool are also experimenting with. The arrangement of the court around an oval or circular table reflects the fact that rather than being designed to express societal revulsion at criminal conduct the surroundings reflect an ideology in which the defendant and other stakeholders are given space to discuss the rehabilitation of the offender.[8]

By way of contrast, Burroughs (2000) suggests that the space under trees was a more appropriate arrangement for courts in monarchical societies where formal power stemmed from a central point and the boughs of the tree were seen as symbolising the protection afforded to the populace by the monarch.[9] Unlike the tribal circle which symbolised dispersed power the tree can be seen as representing the location of judicial power in just one individual. In their account of the building of the Haifa courthouse Chyutin Architects (2006) recount how the prophet Samuel gave the Children of Israel a detailed warning about changes in the law that the transition to a monarchic regime would bring. His predictions were later realised in architectural terms when legal proceedings moved from the edge of the city where people regularly gathered to the palace of the King. The royal hall of judgment, constructed to glorify the institution of the monarch, was panelled with cedar wood and the King was separated from litigants by six steps leading up to a dais.

In some cases trials were held outside because it was impossible or inconvenient to do otherwise. Trial by combat or wager by battle, introduced into English common law after the Norman Conquest, is one such example. Moreover, the trials of crown pleas of felony by ordeal, overseen by the Church in England could not have been conveniently held in an indoor location relying as they did on the use of large quantities of water and fire. Kerr *et al* (1992) contend that there is evidence that ordeal of cold water involved a pool 12 feet in depth and 20 feet in width. Exposure to the elements could be critical to proceedings in other ways. Bellott (1922) has demonstrated that from the first, ancient assemblies such as the Velmic tribunals of Westphalia were intimately connected with the religion of their race with frequent reference being made to the expectation that the court will be held in the eye of the light and face of the sun. Similarly, it was the alignments between Stonehenge and 12 major solar and lunar events that rendered this particular seat of judgment special for those using it (see further Hawkins 1982).

It has been argued that the history of the trial can be seen in terms of the movement from the primitive purity of trials in which God or Gods were deemed to have determined the outcome to the instability of a trial based on precedents and evidence. The purpose of trial by battle or ordeal was to reveal God's will and proximity to the elements may well have reinforced the insignificance of humankind in the cosmos.[10] Similarly, the early system of oaths in which supporters would assert the truth of a litigant's case was founded on the notion that it was only the foolish that would lie about such matters before God.

The English trial has since become more dependent on the secular text or methodology of legal science but whilst God is no longer 'present' in the courthouse in the ways conceived of in earlier times, subliminal religious symbolism continues to remind participants of the gravity of the occasion (Douzinas and Warrington 1991). Some commentators have gone so far as to suggest that the modern courthouse is reified within our culture and best understood as a secular cathedral. There is nowhere where this message is better conveyed than in the cathedral-like great hall of the Royal Courts of Justice (1874–1882) designed by an architect who specialised in ecclesiastical buildings and who believed that architecture must be considered in her religious aspect more than any other (Brownlee 1984).[11] The fact that 'speaking walls' and religious artwork of modern courthouses continue to inscribe the fabric of the courthouse with religious tracts reflects the ongoing association of religion and law.[12] Religious symbolism is also apparent in more mundane aspects of modern court design. In his work on law, culture and ritual, Chase (2005) talks of the sacral architecture of the contemporary court; the altar-like bench, the choir-like jury box, the lectern-like witness stand and the rood screen separating the inner and outer segments of the room.

20 Legal Architecture: Justice, due process and the place of law

It remained quite common for courts to be at least partly open to the elements in England until relatively recently although this appears to have been motivated as much by concerns about disease as anything else. A new Old Bailey sessions house was constructed in London in 1539 but by 1543 the building had been let out and the judges were holding courts in the adjoining gardens. A subsequent court on the site (1668–73) dedicated space on the ground floor to the court but it remained open to the weather on one side.[13] Prior to the fifteenth century it is also the case that local courts were often held at market crosses, constructed to denote the heart of the market place.[14] Over time the simple market cross transformed into a formal structure with posts which held up a canopy to protect goods such as butter from the sun. The fact that the open space below the canopy was also used for courts is evident, for instance, from records of Chelmsford market cross built in 1569. At Rochester Guildhall (1687) the first-floor council chamber doubled up as a courtroom and the other Assize court was held in the open market space below and this was an arrangement which was to become common (Graham 2003).[15] Apart from the protection afforded by the private buildings on either side, courts continued to be open to the elements except when canvass was hung (Chalklin 1998).

The use of partially open courts continues to this day. In an Australian context this practice has been prompted by political sensitivity to the different concepts of meaningful space which prevail amongst indigenous peoples and those ascended from European settlers. So, for instance, the new courthouse at Port Augusta in South Australia has an outside shelter which allows for sittings with a magistrate out in the open in an effect to respect aboriginal connection to land and open space (Grant 2009). Attempts have also been made to bring the outside into the court at the new Supreme Court being built in Brisbane due to complete in 2011. The Banco court has been positioned on the first floor with a large panel of glass along one wall so as to allow participants in the trial to see the horizon. In addition, a large glass atrium within the secure circulation area has been appended to the building to give users access to plants and natural light.

The concept of justice without walls continues to receive some attention from scholars. Mohr (1999) has argued that despite modern assumptions, legal proceedings have never been completely contained by walls. It is certainly the case that the giving of evidence requires participants to transport their thoughts to other times and places, as does judicial consideration of established legal precedent. The media have also been active in violating the physical boundaries of the courthouse by broadcasting proceedings to television networks and the internet where permitted. In his discussion of the OJ Simpson trial Mohr (1999) reminds us of how successful the media were in de-localising the trial so that the US audience could follow every word of the trial on cable channel court TV without ever stepping into the court. Reflecting on the implications of these developments, he has argued that:

> Not only does the place of the courtroom blend into an undifferentiated and incoherent space, but the time sequence of the trial is cut in an out of real

time, flashbacks, and predictions about the verdict. The trial has no more internal structure, once it is predigested and redigested, than the repeated image from the helicopter of Simpson's car driving interminably around the Los Angeles freeways before his arrest. (p.2)

Discussion of contemporary justice without walls goes beyond media coverage of trials. Use is increasingly being made of live video links to transmit the evidence of a physically remote witness or defendant into the court. The fact that this technology could allow a trial to take place in which none of the participants were physically present in the same room has encouraged some commentators to herald the arrival of the trial in a laptop. Far from being confident that design templates are now settled, ideas of 'legal place' are becoming unsettled once again and the legal rituals, traditionally seen as critical to the trial, marginalised as the physical boundaries of the courtroom are increasingly being violated by new information highways (Mulcahy 2008).

Containment

The use of buildings to house state-sanctioned adjudication of disputes undoubtedly brings a number of benefits. The enclosure of space means that the timetabling of proceedings is no longer subject to the vagaries of the weather. The use of buildings also facilitates the storage and better protection of artifacts which accompany hearings. When one looks beyond these practical concerns it is clear that the construction of a building to house legal proceedings adds an important new dimension to the architectural markers available to denote status, rank and dominant notions of due process. Haour (2005) has argued that walls are a particularly potent symbol of power which have played a major role in control, definition and monitoring of bodies in space. For some writers the enclosure of courts within buildings reflects broader shifts in attitudes towards adjudication and the nature of the authority on which adjudicators sought to draw. Graham (2004) has argued that the trend towards holding courts indoors which was almost complete in England from the thirteenth century,[16] reflected the increasing association of legal procedure with the written word. In her opinion:

No longer could [the hearing] be described as a gathering where a community solved its problems according to unwritten custom. Now it was an event in the life of an institution which perpetrated its authority through its own procedures, and above all through written records. A retreat under cover became inevitable, if only because ink and parchment were awkward to use out of doors. (p.42)

In a similar vein, Douzinas and Warrington (1991) draw attention to the move from speech to writing in the English trial prompted by the slow transfer of

22   Legal Architecture: Justice, due process and the place of law

religion from the public to the private sphere and the growth of literacy. Such transformations were by no means rapid but the growth of a 'legal science' with its emphasis on the legal text rather than divine revelation has been traced back to the twelfth century when the first law schools were established specifically for the purposes of studying ancient manuscripts. From a position in which it was expected that the will of God would reveal itself, through for instance an ordeal by fire, Goodrich (1987) argues that in time it was the text which revealed the wisdom of the deity or their disciples and was treated as a sacred source. It can be surmised that once it was the text which was seen to contain a complete and integrated body of doctrine from which all deductions could be made that natural elements became less important in the process of adjudication and a new type of priest emerged in the form of the lawyer. As legal codes and treatise could not be read by all they needed to be 'philologically reconstructed and handed down by an elite group of juristic exegetes' (Goodrich 1987, p.34).[17]

Thresholds

On the cusp of the outside and inside spaces described in this chapter are thresholds. The notion of the threshold of the court occupies a special place in the history of legal architecture where the outside and inside can not properly be seen as separate realms but part of a dynamic interplay of barriers and passages (Arnheim and Watterson 1966). This boundary between inside and outside often assumes an importance of its own and creates a hinterland between open space and manmade construction. So, for instance, in the tribal cities of Eretz-Israel before the period of the Kings justice was administered at the gate of the city which represented a meeting place between the world of the city and its regulated social frameworks and the unruly countryside beyond (Chyutin 2006).[18] In his work on the Inns of Court Raffield (2004) reinforces the importance of the gate in images and cultures of law and argues that '[I]n the iconography, historiography and literature of law, the gate is a recurrent and resonant symbol . . . an immutable determinant of access to law' (p.50). Legal historians have also charted how church doorways were used to conduct certain types of legal transactions which would be recognised by the courts (see for example Foster 1961–2; Haskins 1948–9). The partially enclosed spaces of stoas or covered colonnades of the Greek agora also served to give formal shape to a piece of open space for public events such as the courts but did not completely enclose it.[19] Kostoff (1995) has labelled such locations as 'soft' examples of architecture; spaces which are capable of absorbing some of the legitimacy of the place to which they are adjoined. It is also the case that transitional spaces not dedicated to law have played a part in giving legitimacy to legal events. Burroughs (2000) draws attention to the indispensable nature of architectural structures which lay between domestic and public space as symbolically charged settings for certain kinds of legal transaction in fifteenth-century medieval Italian cities. A key example was the street portico on the road-facing side of a house which were used to mark betrothals and hold commercial arbitrations in the presence of a notary. In this way he argues that these quasi-sacral architectural features functioned as a place for registration and mobilisation of social and collective memory.

The courthouse can be seen as encompassing a series of thresholds and in modern courts the journey from street to courtroom provides participants with visual prompts that reinforce this. Courts are commonly set back from busy roads in order to create a sense of a transition zone from the public space of the highway to the rarefied interior of the courthouse. In his study of Wood Green Crown Court, Rock referred to this as a kind of 'frontier' space in which those approaching the court are forced to make a crossing, to quit the everyday world as they approach the tension point of the entrance threshold (Rock 1993).[20] This aspiration is reflected in the advice to architects provided by the current *Court Standards and Design Guide* (Her Majesty's Courts Service 2010):

> The main entrance and entrance hall require a civic presence to reflect the status of Law in society and engender respect for the decisions made in the courts. This can be achieved by being the focus of the townscape, through symmetry and formality in the architecture, through a generous use of space and height internally and by the use of steps to the entrance ... There should also be a generous external gathering space outside the entrance. The main entrance should symbolically be the image of the court, and the place outside which the Press photograph those seeking publicity after a case (p.1:6.18).

The Israeli Supreme Court provides a particularly striking example of how the importance of the journey to court is used to encourage a transformation in the attitude of the visitor. In his description of the journey to the court, Sharon has suggested that:

> ... one experiences a momentary doubt: where am I heading? The eye, far swifter than the foot, hesitates, bewildered. Perhaps there is an instant of 'alienation' and, at the same time, of reflection. This is a turning point: either one retraces one's footsteps and exits, or one begins, like the space itself, to move toward the courtrooms.

In contrast to the approach prescribed by the *Court Standards and Design Guide* (Her Majesty's Courts Service 2010), the architects, Ram Karmi and Ada Karmi-Melamede, deliberately built an unobtrusive entrance from which a sweeping staircase leads to a large window with a panoramic view of Jerusalem. From there one has to turn to progress towards an inner 'gate' and the courts. The purpose of their design was to create a static and serene space which detached the sacred space in which the courtrooms are housed from the profane entrance which acts as a threshold to normality (Sharon 1993). The turning point at the top of the stairs has deliberately been constructed as a kind of in-between space used to

24 Legal Architecture: Justice, due process and the place of law

engender a sense of gradual transition from the random freedom of the outside to the meticulous hierarchy within the courts

Shared public spaces

Fundamental to the history of ideas about courthouse design is an understanding that despite the proliferation of buildings with legal functions over time for most of the history there has been no such thing as a 'court house', only buildings that happened to house courts. Whilst we have come to imagine the law court as a specialist building type in recent decades with dedicated furniture, fittings and circulation routes designed strictly for legal process, for many centuries the buildings which housed courts were also used for a multitude of other functions. This severely limited the scope of architects to design buildings devoted exclusively and permanently to law. The absence of a specialist building type to house the courts is reflected most clearly in nomenclature. An analysis of Graham's gazetteer of English courts to 1914 shows that the Assizes were held in shire halls, county halls, town halls, guildhalls, castles, market halls, or council houses.[21] Even when they were given the name 'court house' there is often evidence to suggest that they also housed markets or debating chambers. Reflecting on this point, Graham (2004) notes that: 'the very names given to these buildings by contemporaries still retained a domestic quality . . . and were often used interchangeably, suggesting something of [a] confusion of identity'. The suggestion that such confusion existed is to misunderstand the social history of the English. It is much more illuminating to see such arrangements as reflecting different conceptions of public space from those we take for granted today.

In her discussions of the court of King's Bench in the fifteenth and sixteenth centuries Blatcher (1978) argues that the fact that three of the King's courts shared Westminster Hall had an important symbolic and practical significance. The process of sharing is seen as reflecting the indivisibility of the King's justice but also encouraged an interchange of views and information by those involved in the proceedings. She further argues that for men who had passed their schooldays in one small hall where all classes were taught together it was probably not considered a hardship to conduct legal proceedings against the background of other dialogue. It is not just the law court which did not have its own building before the eighteenth century, hardly any specialist or sole use building types existed. It is in this context that Tittler (1991) provides the following account of the town hall from 1500–1640:

> Here the towns [people] carried out their public business. Here their administrative officers normally presided, their courts and assemblies normally convened, their fees and fines were normally rendered, their apprenticeships and freedoms were registered and, in short, their perquisites of government, however defined in particular communities, came chiefly to be exercised.

Here too, were ordinarily kept the paraphernalia associated with such activities: the town mace, seal and plate, town chest and records, the characteristic furnishings of the courts and assemblies held within. (p.9.)

Law courts' spatial association with the market hall, manor houses, shire and county halls and even churches[22] was perfectly natural at a time when the population was smaller and centralised administration of the kind we now experience almost inconceivable. The same space within buildings might variously be used for Assizes, Quarter Sessions and county courts as well as political debates, public meetings, civic ceremonies, feasts, musical entertainment and dancing.[23] The fact that it was the County rather than the monarch or centralised state which was responsible for the provision of buildings for the Assize until the 1970s also made it much more likely that buildings paid for by local residents would have multiple functions associated with local government. As a result, accounts of the architecture of law courts are inextricably bound up with the architectural history of other building types. Regardless of whether towns and their economy were governed by municipal corporations, guilds or manorial courts one building could often accommodate a number of functions and might be paid for by any number of means in the interests of economy.[24]

Shared spaces were also deemed appropriate when the same public officials performed executive, judicial and legislative functions. So, for instance, for many centuries a special jury of the manorial court organised the day-to-day administration of the local market in the marketplace,[25] and the local Justices of the Peace did most of the work of local government at the Quarter sessions including the setting of the poor rate. For the last three hundred years the Assizes were an almost exclusively judicial tribunal but too much emphasis on the Assizes as a court of law conceals the important administrative and political functions it has performed up until the eighteenth century. To the Elizabethan and Stuart governments in particular the distinctive and most important functions of the roving Assizes judges as they travelled between the provinces and Westminster was their role as a vehicle of political and executive control (Cockburn 1972). As Cockburn (1972) explains in his authoritative account of the history of the Assizes:

> Manipulation of the religious climate, the encouragement of political allegiance, efficient local government, the institution and oversight of economic programmes, regulation of enclosure, rating and public works – all fell, or were maneuvered within the competence of the circuit judges . . . In this the court was unique. Not even Star chamber, for all its innovative characteristics and concern with provincial suits, could claim competence in so many diverse fields or such a specific and detailed rapport with the central executive. (p.154)

During this period the Assize judges oversaw the activity of local Justices of the Peace, sheriffs, bailiffs and jurors, bridge repairs, disputed elections, complaints against attorneys, suppression of alehouses and settlement of paupers. As well as reporting back to the monarch's council on these matters they were charged with conveying to provincial agencies the content and implications of central policies and for attempting to ensure that they were enacted. Mechanisms such as the Assize address, which was delivered at the opening of each session provided formal opportunities to reinforce the importance of allegiances to the Crown. There were also numerous informal means to command loyalty and gather local intelligence. Far from being events dedicated solely to pleasure the various Assizes dinners organised for the judges whilst on circuit were frequently used as information gathering events about local politics and the economy (Cockburn 1972).

Given the range of legal systems which operated in parallel with each other up until the nineteenth century the type of building in which courts were held could also depend on the jurisdiction of the court. In hamlets and villages manorial courts were commonly in the large hall at the heart of the manor where people would also sleep, eat and be entertained. Until the *Magna Carta* required royal justice to be dispensed in fixed places the royal courts adjudicated at any number of manors, country estates and castles at which the monarch stayed. The consistory courts which formed the first tier of the system of ecclesiastical courts would commonly hold hearings in a local church whereas the Bishops court would commonly meet in the cathedral of the Diocese.

It is also the case that particular building types have tended to share their history with that of the Assizes. The first of these is the castle. In her gazetteer of court buildings Graham makes mention of 22 castles which have housed the Assizes.[26] As the Middle Ages progressed castles tended to lose their military significance, but a number of them continued as administrative centres and to house courts.[27] Castle defences were often allowed to decay but in some instances the Assizes and other courts went on sitting in the great hall of the castle and in others new shire halls were constructed within the precincts of the castle.[28] When castles deteriorated or fell into private hands the Assizes commonly moved to other locations set aside for community use. The most prominent of these were the multi-purpose market crosses and canopies which slowly came to be known as town halls as they became more enclosed. Graham's gazetteer lists 27 market hall types which were used for the Assizes alone.[29]

The market place was a natural location for courts of law. Not only was it accessible and attracted people but the judicial function was closely associated with local government and local government was closely associated with the market place.[30] The association of law and commerce is apparent from the widespread use of guildhalls for adjudication. It was often the wealthy Guilds,[31] endowed with land and property from bequests, that were often able to provide the most commodious meeting places for important community events.[32] Buildings used to house the Assizes were being given the label guildhall as late as 1885[33] and at Worcester the thirteenth-century Guildhall built to act as a meeting place for Worcester merchants later became the centre of civic administration and retained the role long after the merchant guild had disbanded. Gloucestershire Boothall,[34] dating from 1192, also became the central seat of government in the town and was used for the hundred court, Assizes, markets, travelling shows, concerts and plays (The City of Gloucester 1988).

The spaces provided for courts within these buildings were also shared between them. At the central courts in Westminster Hall three of the courts continued to hold courts at the same time until the early nineteenth century. Guildford court-house (1789) provides a late regional example of a market hall type in which a single hall accommodated the two Assize courts sitting simultaneously at either end of the hall (Graham 2004). In time, such arrangements came to be considered impractical by Assize judges. Girouard (1990) tells the story of the fury of a Crown court Judge at Gloucester where a murder case was being heard and the jury in his court laughed out loud at a joke being told in the Nisi Prius court at the other end of the hall. Competitions between towns for the holding of Assizes courts also made local dignitaries sensitive to the comfort of judges and was one of the factors that led to the enclosure of open spaces below market and town halls.[35] These developments led to the provision of separate rooms for different courts, but commitment to the notion of a single multi-purpose hall which dominated the building was remarkably long lived. So, for instance one immediate response to the problem of noise from other courts in a shared hall interrupting proceedings was not to partition the space but rather to create an 'L' shaped hall as a variation on the traditional rectangle. This was still a unified space which allowed for free circulation of spectators between courts but reduced disturbance by allowing one court to sit around the corner from the other. The 'L' shaped hall made its first appearance at Nottingham[36] (1618). A hall had stood on the same site in 1375–6 but in 1618 a second court was added with the effect that the building was converted into an 'L' shape (Graham 2004). A more deliberate attempt to create an 'L' shaped building seems to have been adopted at Northampton Sessions House (1657–9) which has since been described as 'one of the most architecturally sophisticated courthouses of the century' (SAVE 2004, pp.118–19). The design also makes an appearance at Thomas Harris's courts at Aylesbury (1723–40). However, it did not have a lasting influence on courthouse design and Graham (2003) suggests that this was largely due to the fact that it forced a rather unusual and convoluted ceremonial route on the judiciary.

Ritual and place

Where a chosen site for adjudication has no extraordinary characteristics which mark it out as legal space it is often ritual rather than location which render the proceedings significant. In these circumstances ritual allows the special place of law to be constructed in the minds of those present. Creating a sense of special place through ceremony is particularly important in societies with a system of peripatetic rulership in which justice is administered wherever the ruler happens to be. Rulers who moved from place to place were common in early medieval

28 Legal Architecture: Justice, due process and the place of law

Europe (Haour 2005) and the modern day legal circuits in England are a vestige of a time when the King, and then his representatives, travelled the country hearing law suits as they went. Rituals used to designate a place of adjudication special might be as simple as the act of sticking a pole into the ground or the placing of a spear behind the judge's chair to symbolise the authority of the court (Bellott 1922).[37] But ritualistic practice can also have symbolic meanings which draw on concepts of heritage, divine power, community and nature. At the ancient Free Field Court of Corbey the first ritual to be performed was the 'laying out' of the place of judgment. This involved the measuring of a plot of ground in any open field to ensure that it was 16 feet square. Once this had been done a hole was dug into which each of the Echevins[38] would throw a handful of ashes, a coal and a tile. If doubt later arose as to whether the place of judgment had been properly 'consecrated' a search could be made for these tokens. If they were absent all judgments rendered on that spot were treated as void. A stool, known as the King's stool, was then placed on this spot and a Frohner, or summoner, asked to have his mete wand, or measuring stick, tested in order to confirm that the place of judgment had been measured out properly. It was only after this that the Graff, or senior member of the assembly, would sit on the stool and open proceedings (Bellott 1922; Scott 1831).

The pageantry surrounding the arrival of itinerant Assize judges from medieval times until the 1970s provides a more contemporary example of how ceremony can render an event special. In his account of English towns up until the nineteenth century Girouard (1990) explains how the county sheriff drove out in his coach accompanied by mounted trumpeters, javelin men, Justices and county gentry in order to meet the visiting Judges at the county boundary and how they escorted the judges to their lodgings through streets lined with spectators. The judges were taken to the church for the Assize service and on their daily visits to the courts with equal pomp and ceremony. Graham (2004) charts how in the nineteenth century county policemen began to replace the more colourful javelin men and it became more common for judges to use trains and make their own way to their lodgings. Despite this, Denman's Digest for 1912 reports a decision taken by the circuit judges as recently as 1904 that the sheriff's trumpeters should not be dispensed with:

> There can be no doubt in the minds of those familiar with the subject that the time honoured institution of heralding the approach of the Kings representative by the sound of trumpets has a tremendous effect in impressing the populace with the importance of the proceedings. The day of the abolition of trumpeters would be an evil one.
>
> (Reproduced in MacKinnon, 1940, p.6.)

Reflecting on his experience of being a circuit judge until the Assizes were abolished by The Courts Act 1971, Sir Basil Nield (1972) records that until the abolition judges continued to be met off their reserved carriage in a train by

An ideal type? Visions of the courthouse over time 29

a formally attired High Sheriff of the County and taken to their lodgings. He details the ceremonial opening of the Assize was preceded by a special service in the parish church in which the judges were 'appropriately reminded' (p.6) of the strict and solemn account which they must one day give before the Judgment Seat. Moreover, in a ceremony dating back to the Middle Ages the start of the legal year continues to be marked in London by a procession of judges arriving at Westminster Abbey from the Royal Courts of Justice in The Strand for a religious service, followed by the Lord Chancellor's 'breakfast' in the Great Hall in the Houses of Parliament.[39] The modern day rituals associated with the opening of the law term together with court dress and ceremony in the trial remind us that buildings alone are not always the only signifiers used to denote a special place for adjudication. Instead this has commonly been achieved by marrying both environment and ceremony.

The pageantry of English law surrounding regional courts is undoubtedly less lavish than it once was and it is significant in the current context that ceremony surrounding the courts appears to have diminished as purpose-built law courts with their permanent emblems of justice became fashionable. It would seem that in times when it was common for the Assize judges to share accommodation in buildings constructed for a multitude of purposes associated with local adminis-tration, ceremony played a crucial role in rendering a local public place a legal place in which justice was administered in the name of the centralised power of the monarch. As more elaborate and distinctive purpose-built courthouses were constructed it could be argued that it was architecture which was increasingly employed to mark a location out as a special place.

Parallel functions

Ceremony and ritual remain incredibly important to the administration of justice today but their centrality to proceedings has been undermined as architecture adopted the role of marking out specific spaces as dedicated to law. A deliberate shift away from the provision of shared legal space was first heralded by the building of the Guildhall in Worcester (1721–3) in which the two Assize courts were placed behind the main hall. Even then there was not complete separation as the courts were divided off from the other public spaces by arches rather than walls. A similar trend first identified at Abingdon County Hall (1677–82) shown in Figure 2.2, was to separate the courts from the hall by way of a gallery sup-ported by columns and this use of columns with or without a balcony became a common way of marking courts off from communal space (Graham 2004). Significantly, this arrangement allowed for a degree of privacy without a com-plete separation from the great hall and was even used to separate the two Assize courts where there was no communal hall.[40] The Shire Hall at Warwick (1753–8) had a similar plan which Girouard (1980) suggests was inspired by Worcester and arrangements in which there was symbolic rather than actual separation from the main hall were also adopted at Bedford Shire Hall (1753), Nottingham (1768–72),

30 Legal Architecture: Justice, due process and the place of law

Figure 2.2 Abingdon County Hall (1677–82) © Linda Mulcahy.

An ideal type? Visions of the courthouse over time   31

Exeter (1773) and York (1773–7) where the courts were positioned at either end of the hall but separated from it by pillars (Graham 2004).

Although pillars and arches signified a separation of space into that designated legal and that designated multi-purpose contemporary architects and those commissioning work appear to have remained concerned about facilitating an open court. With only arches or pillars to separate out legal space from general public space the courts remained accessible to large numbers of people and wandering spectators could enter and leave the court with relative ease. Moreover, the spectator was not confined to those who sat or stood in the body of the court but could also be used to describe those in the vicinity of the court. Whilst they might not be able to see or hear the proceedings this group could gleen fresh intelligence from those who could and in the very least could ascertain the mood of the court. Those at the back of the adjoining hall might be frustrated at not being able to get closer but there at least remained a semblance of the notion that the administration of justice was a public event.

This hiving off of legal space from other public spaces within shire, town and market halls is probably the most significant stage in the transformation of courtroom design because it reflects changing notions of law as an activity which was different from politics or administration. It is in the late eighteenth-century courts that such formal separation of legal space from communal space first gains in popularity in new build. In his authoritative book on public building projects of this period Chalklin (1998) notes that all of the seven new County halls built between 1768 and 1782 were substantial buildings with separate rooms for crown and civil courts within the building.[41] The increasing popularity of demarcated legal space can also be seen from the alterations made to courts in the eighteenth and nineteenth century. Graham's gazetteer shows that when Winchester Great Hall (1222–36) was remodelled in 1765 the courts within the hall were enclosed by curving partitions. The two courtrooms at Oxford Town Hall which had initially been divided by rows of pillars were replaced by sliding partitions in 1791. Similarly, in 1812 Northampton Sessions House which had, for a short time, led the way in its innovative 'L' shaped design had its Crown court divided off from the hall and a gallery added.[42]

Whilst the practice of constructing courthouses dedicated solely to law became popular from the late eighteenth century onwards it is significant that the practice of shared space remained the dominant model for English law courts across their history and that the idea still occasionally enters into contemporary policy debate. As late as the nineteenth century civic buildings were still being constructed to house law courts alongside other public facilities. Indeed, some of the great buildings of the era adopted this practice. Leeds Town Hall (1853–8) was built to house the council offices and chamber, three courtrooms and associated space for legal personnel, the police station, a grand concert hall and a central library (Leeds Council 2009). In Liverpool, St George's Hall (1847–56) provided accommodation for two Assize courts and two concert halls. As late as 1890 Kingston County Hall was built to serve both local politicians and administrators, as well as the

32 Legal Architecture: Justice, due process and the place of law

judiciary. Whilst in some buildings the court existed in parallel with other facilities, in others the different users continued to share certain spaces with other public bodies. This was particularly true of the space set aside for use as a Nisi Prius court. The absence of the dock in this court meant that local authorities often used the civil Assize court as a debating chamber as well. Indeed, this probably explains the common practice of having curved seating in some courts.[43] An excellent example of how this altered the furnishings and design can be seen at Aylesbury Crown Court which was built in the eighteenth century and is still in use as a Crown court. The nisi prius court doubled up, and continues to be occasionally used, as a debating chamber for local politicians and is much more elegant and inviting than the criminal court which had but one use. The Howard League for Penal Reform's (1976) report on the design of the dock makes clear that this practice of nisi prius courts doubling up as council chambers survived elsewhere until the 1970s. But frustration at having to share space was also evident during the 1970s when the Royal Commission on the Assizes and Quarter Sessions complained that in many venues used for courts the halls and corridors surrounding the courts were often stacked with paraphernalia associated with other uses of the building such as dismantled staging, parts of a boxing ring, or the music stands of a brass band contest.

Today, many policy makers are turning their gaze once more to the idea of a multi-function space. In a US context Green and De Nevi (1989) recount how Frank Lloyd Wright's design for the Marin County Civic Centre became a turning point in the design of courts in the US and sparked a new interest in the subject. The Centre included courts, a post office, a county library, local government offices, a jail, fairground, parks and a playground. More recently still the concept of shared public spaces has taken a new turn as community courts or neighbourhood justice centres are being promoted by policymakers in the United States, United Kingdom and Australia. Aimed at providing resources for the disadvantaged the schemes envisage that the courtroom provides a hub around which all sorts of criminal justice agencies and social services are based. So, for instance, Liverpool Community Justice Centre combines the powers of the courtroom with a range of community resources available free of charge to residents such as victim and witness support, financial advice, housing advice and education and employment services.[44] This suggests that the notion of courthouses and courtrooms having fixed spatial identities continues to be contested.

A distinct domain

Once partitioned 'courtrooms', as they had finally become, were much less likely to serve a multiple function. Communal multi-purpose spaces continued to exist within the buildings which housed courts and to host plays, assemblies, balls and public meetings but spaces dedicated to the administration of justice operated in parallel with these. Significantly, the creation of a separate zone for the courtroom allowed those who oversaw the trial to develop more sophisticated notions of legal space. If fixtures and fittings no longer had to be portable or used for multiple purposes then it became worth designers' while to give more attention to how space could be used to reflect a room's specifically legal purpose. Court officials and planners embraced this task with considerable enthusiasm. The result was that the internal configurations of courtrooms became progressively specialised. In the environs of the court rooms for administrators, waiting witnesses, subterranean routes to the court and dock and separate access from the street began to emerge. Within the court more extensive barriers between the various participants in the trial became common. An excellent example of these developments is the Shire Hall at Ely (1821–2) which has survived with its original joinery intact. A succession of wooden barriers separate the space within the criminal courtroom into a series of separate segments for judges, witness, jury, defendant and public.

As the population increased in size and English society became urbanised so too did the taste for specialised public buildings. Law was no exception and the rise in litigation soon led to a trend for purpose-built courthouses in addition to purpose-built courtrooms. Stevens (2002) argues that these were unknown before the eighteenth century but SAVE (2004) list Derby Shire Hall (1659) as the earliest surviving building which was purpose-built to house the Assize courts[45] and the Old Bailey of 1668 is another early example of an early courthouse solely devoted to adjudication. Graham's (2004) gazetteer makes clear that the proportion of sole use purpose-built courthouses remained low in the eighteenth century but in the nineteenth century 18 of the 40 courts built for the Assizes were purpose-built and sole use. Early examples include Thomas Rogers' Middlesex county sessions house (1779–82); Beverley Sessions Court (1805) and Carlisle Assize courts (1808). In the early years of their development these buildings appear to have been dominated by the courtroom. John Carr's elaborate Assize courts at York (1773) and John Smirke's more austere courts at Maidstone (1824) provide excellent exemplars of early designs for sole-use courthouses which demonstrate that the majority of floorspace within these buildings was dedicated to spaces for adjudication. In many ways, the first purpose-built courts reflect the arrangement of the partitioned great halls discussed earlier in this chapter in which a central entrance hall was flanked on either side by a courtroom. Separate rooms to house administrators and waiting rooms were located on the periphery. As the purpose-built law court gained in popularity in the nineteenth century the internal configurations of the courthouse changed drastically and increasing emphasis was placed on herding different categories of court user around the building in contained circulation routes. It is in these Victorian designs that we begin to see the emergence of the modern courthouse of the type which might be recognised by Stephen Quinlan's critic and it is to these palaces of justice and their segmentation practices which we turn in the next three chapters.

Conclusion

In this chapter I have attempted to map out the many spatial practices associated with courts and adjudication which have existed, and perhaps most importantly co-existed, over time. My aim has been to present this account as a history of recurrent ideas and concepts rather than as a chronology of courthouse design which suggests that the architecture of courts can be understood as a progressive trajectory in which each stage improves on the design of the last. I have suggested that debates about justice without walls and the notions of the law court as a generalised rather than a specialised public space continue to haunt debate. What the analysis offered in this book hopes to make clear is that the one feature which best characterises the history of court design is its status as a multi-purpose public space with a lack of a specific identity as a house dedicated to law. These are all themes which will continue to be pursued throughout the book.

Notes

1 In Book 18 lines 478–608 of the *Iliad* Homer (1950) gives a detailed description of the imagery on the shield decorated by Hephaestus. A series of opposing scenarios such as the city at peace and the city at war demonstrate the ideal forms of a civilised society on earth.
2 While a number of moot hills are naturally occurring some were originally created as burial mounds and others drew their significance from the rituals performed there. This is true of the Tynwald Hill court on the Isle of Man; an artificial mound which was constructed from soil taken from the 17 parishes of the Island. See further http://www.tynwald.org.im/ (accessed January 2009).
3 See for instance British History online (2009, a, b,c,d,e,f,g,h,i) and Mulholland (2003).
4 British History online (2009j).
5 See further http://www.constitutionalcourt.org.za/site/thecourt/thebuilding.htm (accessed January 2010).
6 In his work on the Greek temple Scully (1993) reminds us that the preponderance of open air altars well after the arrival of the temple and enclosure of sacred images indicates that the site on which worship occurred remained important and first suggested the presence of a God. Even when the temple gained popularity the natural environment was still used to bring the design and experience to a climax. So, for example, Vitruvius later argued that temples should always face east so that the worshipper could see the Gods approaching. Moreover, Gods and Goddesses often took on local names from the site of their temples such as Hera of the Cliffs.
7 Assemblies of free neighbours or folk moots held across Western Europe in the early Middle Ages decided disputes according to tribal or regional customs.
8 Spatial arrangements alone may not be enough to promote this ideal. See for instance Griffiths and Kandel (2009).
9 See further Burroughs (2000); Jacob (1995–6); Homer (reprinted 2003) and *Judges* 4.
10 See further Lee (1984).
11 Street was the architect of 179 ecclesiastical buildings.
12 So for instance the artwork under the dome of the Old Bailey includes images of Moses, Solomon and John the Baptist. In its discussion of the mottos on the walls of Leeds Town Hall which contained courts *The Leeds Mercury* (1858) declared that they were 'durable mementoes of the first truths of religion and virtue, the great sentiments of political liberty and the best maxims of commercial industry'.
13 One reason for allowing fresh air to circulate was fear of gaol fever. In Oxford in 1577 a two-storied sessions house within the castle bailey was abandoned after the notorious 'Black Assize' at which over 300 people died of gaol fever including the judge and jury (SAVE 2004).

An ideal type? Visions of the courthouse over time   35

14 From the fifteenth century onwards it became common for these constructions to evolve into covered markets with rooms above. During this period these buildings often continued to be called market crosses but also came to be known as market halls or town halls (Girouard 1990). The association of law courts with market crosses continued to be reflected in nomenclature long after courts sat in market halls. So for instance the court building in Chelmsford constructed in 1569 was called the 'Market Cross'.

15 In some instances the market hall only had a roofed structure on columns. This was the case for instance with Chelmsford Market Cross (1569). The nisi prius court was not enclosed until 1714. The Crown court was reconstructed in 1709 but remained open to the weather (Graham 2004).

16 Graham (2004) has suggested that by the thirteenth century an open air setting had become exceptional for the twice-yearly Shire courts. By the eleventh century England had become a single kingdom with a more sophisticated system of administration organised around the shire or county unit. Twice a year the ealdorman and local bishop assembled the local shire to raise taxes and deal with any other administrative matters including judicial business. Shires also had hundred courts and manorial courts which met more regularly.

17 A major premise of the common law which can be traced to this period is that trials are processes in which pleaders speak their cause in the presence of one another and that it is through discussion and argument that the truth of the cause will emerge (Douzinas and Warrington 1991).

18 See further *The Bible*, Deuteronomy 16.18; Psalms 118.19; Proverbs 22.22.

19 These offered shelter from the weather but were generally open to the elements on three sides. By the fifth century BC these secular constructions had developed from sanctuaries into locations for public sessions of the courts, banquets, displays of public art and places where public notices could be displayed. By way of contrast Byzantine stoas were covered halls. (Kostoff 1995).

20 Within the courthouse, the psychological power of the 'shadow' of the courtroom is all too familiar to socio-legal scholars who still talk of lawyers bargaining at the door of the court. Part of the practical value of the modern *salle de pas perdu* adjoining the courtroom is that it provides a space in which last minute negotiations can take place. The importance of this territory is demonstrated by the complaints that lawyers have lodged over time about courthouses which provide insufficient space in the environs of the court for such private discussions to take place. The lack of space immediately outside the courts in which consultation and negotiation could take place was one of the criticisms lodged against the Royal Courts of Justice when they opened in the Strand.

21 An analysis of Graham's Gazetteer of buildings built to house English courts from the twelfth century onwards demonstrates that of the 119 which housed the Assizes up until the building of Manchester Assize court, 38 were known as shire or county halls, 20 as town halls, 14 as guildhalls, 13 as castles, 5 as market halls, 5 as halls, and 2 as council houses. The remaining 22 were known as courthouses. See Graham (2004).

22 The Cannock and Rugeley court was reported to have been used for the manor court as it was the only place in the vicinity big enough to house it. See further chapter four in Brookes and Lobban (1997).

23 For a colourful account of the various uses of Gloucester Boothall see British History online (2009k).

24 Some buildings such as Reigate Town Hall, Horsham Town Hall or Appleby courthouse were paid for by a rich benefactor. Others such as Northampton sessions house and Maidstone Town Hall were funded by the borough and county. Devizes Assize court was paid for by public subscription. See further Chalklin (1998).

25 Girouard (1990) suggests that this practice can be traced from the medieval era to the nineteenth century in some towns.

36 Legal Architecture: Justice, due process and the place of law

26 These are Appleby, Cambridge, Canterbury, Chester, Exeter, Guildford, Hereford, Hertford, Lancaster, Launceston, Leicester, Lincoln, Newcastle, Northampton, Norwich, Oakham, Oxford, Shrewsbury, Taunton, Winchester, Wisbech and York.

27 Oakham Castle great hall, which has been the site of a court since 1229, still hosts an occasional Magistrates' court and the Assizes were held there until 1970. Lancaster Castle which first held Assizes in 1165 is still in use as a Crown court and the Crown and County courts have only moved out of Exeter Castle this decade.

28 Graham dates this trend from the thirteenth century onwards. Examples include Cambridge Shire Hall (1572), Norwich Shire Hall (1747–49), the Crown court at Chester Shire Hall (1788–1822), Newcastle moot hall (1810–12), Norwich Shire Hall (1822–4) and Lincoln County Hall (1822–8) in which the Assizes have been held uninterrupted on the castle site since 1068.

29 Examples of the latter are evident in a crop of buildings constructed in the seventeenth century with examples at Maidstone (1608), Hertford (1610), Abingdon (1677) and Southwark (1685).

30 Girouard (1990) notes that in 1600 there were in the region of 800 market towns and that there had been many more in the Middle Ages. He argues that only the church rivalled the market in terms of attracting crowds. The market had to be regulated; rents collected, a weigh beam provided and overseen, and disputes on market day settled. Disputes on market day were often settled speedily by a Court of Pie Powder. Pie Powder is a corruption of Old French *pied poudre* meaning 'dusty of foot'. This referred to travelling traders who were thus instantly recognisable as strangers or alien merchants. See further Gross (1906).

31 Merchant Guilds were associations of merchants recognised by the Crown and set up from the twelfth century. Most medieval towns had several and one usually became especially powerful. Many Guilds were dissolved by Edward VI in 1549 (Girouard 1990).

32 On occasions, where powerful Guilds, Corporations and Manors existed side by side these authorities might actually share the costs of a building. Guildhalls were used to house courts in a number of locations including Bath (1551–2), Exeter (c1160), Guildford (1589), Leicester (1390), Lincoln (1160), Newcastle (1655), Norwich (1407–53), Rochester (1540), Thetford (1799) and York (1449–53).

33 Nottingham Guildhall was built 1885–8.

34 Boothall is used interchangeably with Guildhall at Gloucester and elsewhere.

35 By way of example Horsham Town Hall (1721) had the ground floor boarded in to create a distinctive area for the Criminal Assize courts in 1809.

36 Graham (2004) notes in her Gazetteer that Shrewsbury Shire Hall was an L shaped timber building, but it is not clear whether this design was adopted in order to accommodate the two courts.

37 Homer's description of the stone circle on Achilles' shield makes clear that when the elders wished to pronounce judgment they took a staff from the herald to signify this.

38 These appeared to be a type of alderman.

39 See further http://www.judiciary.gov.uk/publications_media/media_releases/2009/2509.htm (accessed January 2010).

40 An example is Oxford Town Hall 1751–3.

41 These were the courthouses at Hertford, Lincoln, Nottingham, York, Exeter, Appleby and the Middlesex sessions house at Clerkenwell.

42 Other examples from Graham's (2004) gazetteer include Leicester Castle where the hall was divided into two courtrooms in 1821; Bedford Shire Hall where the courts were partitioned off from a central hall in 1831.

43 At Chester Shire Hall (1788–1822) the curved shire hall was converted into a Crown court.

44 See further http://www.communityjustice.gov.uk/northliverpool/index.htm. The purpose-built Neighbourhood justice centre in Collingwood near Melbourne also

houses mediation facilities, a play area, alcohol and drug addiction services, financial counselling, legal advice, housing support, victim support and a mental health clinician, all of which co-exist with the court.

45 SAVE (2004) also make it clear that it continued to be used for assemblies and meetings outside of the bi-annual visits of the Assize judges.

Chapter 3

Segmentation and segregation

Introduction

In this chapter I look at the origins of current ideas about modern court design. Central to my argument is the observation that as the organisation of society has moved from one based on the ideals of feudalism to those of a representative democracy the arrangement of the courthouse has, somewhat paradoxically, become increasingly hierarchical and discourses of containment and surveillance more pervasive. I attempt to chart the ways in which court architecture has become increasingly segmented and the public space of the courthouse less accessible to lay participants in the trial. Analysis of architectural plans and historical records provide a number of indications of the ways in which changing notions of privacy and due process have prompted the development of complex spatial configurations in the courtroom which can be seen as running counter to debates about the civil liberties of the defendant or the importance of publicity in the trial. In this chapter and the two which follow, it will become clear that current ways of thinking about how and why the interiors of courthouses and courtrooms should be partitioned into zones, and movement within them restricted, have come about as a result of turf wars about who can legitimately participate on the legal stage and the respect which should be afforded them.

I suggest that the containment of the public, the fortification of the dock, the increasing amount of space allocated to advocates, and the creation of dedicated space for journalists all have complex histories which deserve to be charted and discussed more extensively than they have been to date. In contrast to a vision of adjudicatory space as neutral this chapter argues that understanding the geopolitics of the courtroom is crucial to a broader and more nuanced understanding of the history of the trial. The spatial is clearly open to, and a necessary element in, the socio-legal dynamics of the trial. Each time a partition is created it has the effect of creating an inside and outside; an 'opposition' or other which can serve to signal status, place, inequality or sanctuary. Viewed from this perspective, the suggestion that space is fundamental in any exercise of power by ensuring a certain allocation of people in space and a coding of their reciprocal relations is a compelling one (see Foucault 1984; Massey 2005). It would seem then that space is very far from being a flat, immobilised surface. The space-place dynamic is particularly striking when one considers that in the modern court space ostensibly designated 'public' is in fact divided into a series of private spheres which are not accessible to all.

The vulgar and the sacred

The idea of equality before the law is one of the mantras of modernity but a major theme to emerge from historical accounts of the trial is that differentiation between participants has a very long heritage and continues to dominate thinking about court design. In the earliest written account of a trial Homer details the presence of an inner ring of judges and litigants and an outer ring of spectators restrained by a herald. In an English setting, Bellott's (1922) work on the 'bar' traces segregation back to the ancient courts of the Celts and Anglo-Saxons. He describes for instance, how none but the echevins, or aldermen, were allowed in the inner sacred enclosure of the Free Field Court of Corbey. The common practice of 'fencing the court' was an exercise in ordering people and creating an inner and outer area for trials. Indeed in his discussion of two alternative theories of the origin of the word 'bar' Bellott (1922) concludes that the collective name of advocates in the modern trial is associated with the concept of differentiation in which the sacred are separated from vulgar in the precincts of the court (p.183).

The earliest illustrations of an English court suggest that four types of participant in the trial; the judiciary, court officials, litigants and spectators, have long been segregated and that this practice has remained constant until the present day (Graham 2004; see also Jacob 1995–6; Burroughs 2000). The fifteenth century Whaddon illuminations[1] illustrate the layout of the four Westminster-based courts of King's Bench, Exchequer, Common Pleas and Chancery, and demonstrate that like their modern counterparts, the national courts in medieval times were divided into a number of zones. This was achieved by court staff making use of furniture available in public halls for a multitude of other functions. Each of the courts in the Whaddon illustrations has a raised platform occupied by the judiciary seated on a shared bench,[2] a practice which derives from the presence of a raised dais in medieval halls which could be used by those of higher rank. Immediately before the dais is an area in which the court officials sit around a large table which dominated the space and in the case of King's Bench this is shared by the jury who stand at one side of the table. The backs of the three benches placed at the table and on which the various officials sit form the inner 'U' shaped bar of the court beyond which defendants, their lawyers and spectators can not come.[3] Lawyers and their clients occupy an area in the foreground which is in turn separated from spectators by an outer bar which distinguishes the court area from the general space of Westminster Hall. In the King's Bench the outer bar is formed by a raised wooden pole, in the court of Chancery by the back of a bench occupied by law students and in the court of Exchequer by the back of the cage containing those

40 Legal Architecture: Justice, due process and the place of law

waiting for trial. This scheme of ordering people in the court of King's Bench is shown in Figure 3.1.

It can be seen from this illustration that the arrangements creates four categories of user. The judiciary on the raised platform are clearly accorded the

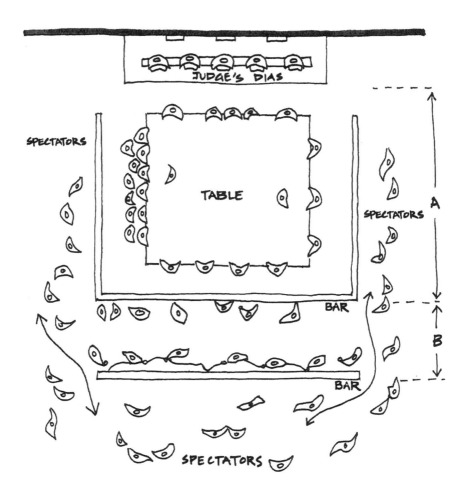

Figure 3.1 The fifteenth century court © Emma Rowden.

highest status, followed by the numerous court officials who occupy the centre of the court. The litigants and their lawyers occupy the third tier and spectators are kept at the margins of proceedings by the outer bar.

Graham (2004) identifies the earliest visual evidence of a courtroom, equipped with more than makeshift furnishings in a multi-function hall, as being a painting of the Court of Wards and Liveries in session at the Palace of Westminster in about 1585. This court, which existed for little over a hundred years, was housed in a specially built extension off Westminster Hall.[4] What is most interesting about the painting for present purpose is that despite the possibilities offered by customised court fittings to experiment with different internal configurations the layout of the court remains fundamentally the same as those illustrated in the Whaddon illuminations. A large table in the centre is used by judges and officials and the backs of the benches the officials use mark the inner 'U' shaped bar behind which litigants and lawyers make their case. The only difference would appear to be that litigants and spectators mingle together in the outer zone of the court. In fact, the arrangement of the court illustrated in these early sources became so embedded that the use of a large table in the well of the court continued well into the nineteenth century even when permanent and purpose-built constructions became the norm. In her analysis of these early paintings Graham (2004) has characterised the arrangement as denoting a court in which the parties in each zone have freedom of movement within them. Indeed, paintings and sketches of courts up until the nineteenth century which she reproduces in her book show courtroom scenes in which spectators promenade between courts and participants in the trial talk amongst themselves. The layout of the consistory court at Chester Cathedral which survives substantially unaltered since 1636 and is shown in Figure 3.2 suggests that the arrangement of the court around a large central table and the fencing of the inner area with high-backed benches was also evident across legal jurisdictions in England (Graham 2004). An illustration of the Earl Marshall court at the College of Arms around 1707 follows this plan with a central table bordered on one side by a raised dais and surrounded on the others by seated officials with a railing behind them. In common with other early illustrations these designs reflect the importance of particular individuals. In the Westminster Court of Wards the official presiding over the court is sitting on a higher-backed bench than other participants with a coat of arms behind them.

The longevity of certain aspects of this template is demonstrated when one compares these arrangements with the standard design for a modern Crown court illustrated in Figure 3.4 (Her Majesty's Courts Service 2010). We continue to adopt the practice of placing the jury on one flank of the courtroom and the court clerk still sits directly in front of the judge with their back to them. The defendant continues to be positioned beyond the officials of the court facing the judge and spectators are still assigned space at the periphery of the court. In addition the nomenclature used in the modern trial whereby we still refer to 'the bench' and the 'bar' reflect the central role of these early signifiers within the legal

Figure 3.2 The consistory court in Chester Cathedral © Linda Mulcahy.

imagination. It would be wrong to assume that the simplicity of the barriers used in these early courts reflected a less sophisticated form of segregation than we are used to in modern courts where custom-built barriers have divided the court into ever-increasing segments. As Graham (2004) reminds us in her discussion of Louis IX's management of French trials in the thirteenth century, societies which are organised in terms of rank rather than profession may require less explicit prompts to distinguish between those who are empowered to act and those who merely observe. So, for instance, the participants in the Whaddon manuscripts are wearing a range of different robes or carrying distinctive staffs which make their rank and status clear and the fact that the prisoners have been incarcerated is all too obvious from the chains around their feet, their poor state of dress and long dishevelled hair.

Recognition of the ways in which certain aspects of the inner space of the courthouse have remained constant is an important starting point for this chapter. It demonstrates that segregation and the imposition of hierarchical ordering has been endemic in the design of the courthouse across time and legal cultures. But of equal importance is the exploration of the changes which have occurred to the internal configuration of the courthouses and courtrooms and what this tells us about the changing jurisprudence of the trial and notions of due process. Of particular interest is the way in which the spatial configurations of the courthouse and courtroom have changed as the practice of law separated itself out as a distinctive

Segmentation and segregation   43

public function which required purpose-built premises and fittings. From this time onwards spatial differentiation in the trial became increasingly pronounced. Such differentiation was achieved in two ways, by segmentation of the public space in which people gathered in the courtroom and by segregation of participants in the environs of it. It is to these two developments that we now turn.

Segmentation

For much of their history buildings used for trials were of simple construction. The practical implications of sharing space were considerable since furniture and props had to be adaptable to a number of different purposes. By way of example, the Guildhall at Exeter, first mentioned in documentary evidence around 1160, had 'court' furnishings which seem to have consisted of just four benches set in a square made of stone slabs filled with earth and sand. Inscriptions on early deeds indicate that these four benches alone defined the area of the court. Similarly, the Guildhall at King's Lynn built in 1422–38 for the Trinity Guild and converted into a town hall in the sixteenth century had moveable court fittings which could be taken away when the room was needed for entertainment (Girouard 1990; SAVE 2004). In short, the trappings of the court had to be simple, easy to dismantle and suitable for a number of uses.

Facilities available for the visiting Assizes commonly consisted of little more than one or two multi-purpose rooms with the occasional ante chamber. It was common in the Middle Ages for both corporations and manorial courts to have just two rooms to conduct all public business. One of these rooms, which was likely to be the largest indoor space in the settlement, would be used for courts and entertainments and a small room off it would be used for everything else.[5] In his work on the English town Girouard (1990) suggests that the authorities in small towns could often manage to organise all of the necessary functions of local government in a single chamber with just one chest for records. He argues that one big room furnished with a raised platform at one end, a table for corporation meetings and banquets and a desk for a clerk satisfied most local needs. The town clerk, who was almost always a practising attorney, had no need for dedicated space within the town hall as they tended to keep their own records and work from home. As was made clear in the last chapter the use of a hall to house more than one court continued in a number of locations until the early nineteenth century.[6]

The simple arrangement of furniture shown in Figure 3.1 began to change in a number of ways from the fifteenth century. From this period onwards there was an increasing segmentation of the court into a number of discrete zones which marked out increasingly differentiated participants in the trial from each other. Figure 3.3 shows how these practices manifested themselves in terms of an increasing number of physical barriers within the court in the courtroom at Presteigne County Hall (1826–9). In some cases newly recognised figures in the trial needed to be assigned places in the courtroom. This is true for instance of the witness, a figure which was practically unknown in jury trials prior to the 1400s

Figure 3.3 Segmentation at Presteigne County Hall (1826–9) © Linda Mulcahy.

and not common until the 1500s. Prior to the recognition of witnesses members of the petty jury had fulfilled the role of both trier and witness. Their personal knowledge of the events in question was a chief source of information which is today furnished by ordinary witnesses (Studnicki and Apol 2002). Remaining evidence of places set aside for witnesses suggests that they could be placed anywhere beyond the inner bar of the court. The lone raised seat placed outside of the inner 'U' shaped bar of the consistory court in Chester Cathedral (1636) towards the back of the court which was likely used by witnesses. Illustrations of Winchester Hall from the first half of the eighteenth century also show a simple marked-out area for witnesses to stand in between the jury and the inner bar of the court (Graham 2004). The re-fitting of the court of King's Bench from 1739 and the surviving fittings in Beverley Guildhall (1762–4) saw the placing of witnesses in a much more clearly differentiated box which flanked the central table and was closer to the bench. It is possible that in the noisier proceedings of the eighteenth century that this arrangement better suited the judge who was much more likely to hear the evidence being given if the witness was placed close by to them. At Nottingham Shire Hall (1769–72) the position of the witness box was even moved to the other side of the bench for a resident judge who was deaf in one ear.[7]

Where they had once shared observation space with members of the public, the press were gradually allocated dedicated desks within the body of the court as their constitutional role as potential defenders of the public interest in the trial gained acceptance. The Crown court at Dorchester (1796–7) provides an early example of this in the form of a fold-down desk for reporters in the corner of the public gallery. When the Assize courts at Durham were refurbished in 1870 the press were given a prominent place in the well of the court adjacent to the witness box and judicial bench. In some instances this space was allocated at the expense of that given to members of the public attending the trial. Graham (2004) recounts how when Wyatt constructed an eastern addition to the courts at Winchester in 1871–4 he gave the press ringside seats whilst deliberately restricting the accommodation for spectators.

The Whaddon illuminations show the jury in the fifteenth-century courts of King's Bench standing together on one flank of the central clerk's table but elsewhere such grouping of the early jury seemed less common. Pole (2002) asserts that since its inception in the thirteenth century members of the petty jury often sat about at random in the public areas of the space allocated for trials and did not even withdraw from the court to confer. Illustrations of the interior of the Old Bailey which had been rebuilt in 1674 following the Great Fire of London show partitioned spaces for jurors to both the left and right of the accused and there is evidence that it was only from 1737 that they were brought together in stalls on the defendant's right (Gale 2006). Graham's (2004) claim that even in the eighteenth century the courts were sometimes so crowded that the jury could not find any seats suggests that such grouping of the jury was not universally adopted. What is clear is that by the end of the seventeenth century it became common for

46   Legal Architecture: Justice, due process and the place of law

juries to give their verdict at the end of each case as opposed to the practices of delivering them in batches after a series of short trials and this made it practical for the jury to be given space together within the court so that they could confer.

As the trend towards the adversarial trial grew in momentum it is evident that lawyers also begin to make territorial claims on the court. The large clerk's table which dominated the well of courts until the nineteenth century had originally been occupied by court officials alone. It can be seen from the Whaddon illuminations that even the most prestigious of lawyers were required to stand with their client at the inner bar formed by the backs of the clerks' settles. As lawyers began to play a more active role in the trial, barristers slowly came within the bar to sit at the central table and the central table itself was gradually replaced by a much smaller clerk's table and dedicated rows of seating for barristers and solicitors which faced the judge (Graham 2004). These various factors meant that the spatial configurations of the trial changed from a simple arrangement of a table, four benches and some beams in the fifteenth century to a much more complex arrangement by the end of the eighteenth century in which architectural features were used to denote a broader set of categories of participants in the trial and their status.

Figure 3.4, which illustrates the standard layout of a Crown court as prescribed in the *Court Standards and Design Guide* (Her Majesty's Courts Service 2010) demonstrates just how far court design has changed since the fifteenth century. The Whaddon illuminations suggest that there were four main categories of people which the design of the court helped to distinguish. These were the judiciary, the officers of the court, the litigants assisted by lawyers in the pursuit of their case and the general public. By way of contrast the modern-day court is highly segmented with separate sections of the court being allocated to the judiciary, clerk, usher, jury, witness, counsel, solicitors, defendant and security staff in a way which is much more ordered. The sociability of the fifteenth-century court in which movement was common has been replaced by a much more static space in which a plethora of modern legal actors are allocated a bounded and discrete place.

This trend was not limited to the courtroom itself. Increasing segmentation of the public also proved to be popular in the immediate vicinity of the courtroom and buildings for courts began to be constructed with much more than a courtroom in mind. Court design moved from a position which was common as late as the eighteenth century in which all the functions relating to the hearing of a case were carried out in one hall to a situation in which courtrooms in the twentieth century commonly occupied just 10 per cent of the total space in courthouses (Jeavons 1992). In one sense the growth of rooms in the vicinity of newly built courts of the late eighteenth century represented a bringing together of justice facilities under one roof. There had always been offices and locations outside of the courthouse associated with its business. Many of the records of the court might be held by a local attorney at their offices and consultations with clients or meetings of jurors would often occur in the local inn or coffee house. In her work on the court of common pleas in the fifteenth century Margaret Hastings (1947)

Segmentation and segregation 47

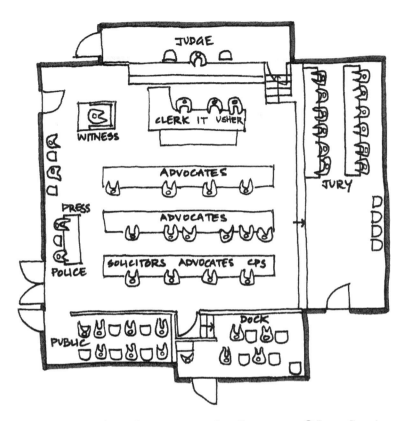

Figure 3.4 The layout of a modern courtroom in a Crown court © Emma Rowden.

argues that there is evidence that the administrative work of the courts was often conducted in churches in West London and towards the end of the fifteenth century court officers were beginning to set up offices in the Inns of Court. The rationalisation of these facilities under one roof made clear that from the late eighteenth century the administration of justice was increasingly associated with one location.

The separating out of space dedicated exclusively to law touched upon in the last chapter provided architects and designers, judges and court staff with important new opportunities to claim territory which was adjacent to courtrooms. Discrete zones for the courts and accompanying accommodation became common in locations shared with the town hall and the idea of law needing a separate and discrete space was finally fully realised as the sole use courthouse gained in popularity. Chalklin (1998) identifies seven new Assize courts built between 1768–71 at Hertford, Lincoln, Nottingham, York, Exeter, Appleby and Middlesex

48   Legal Architecture: Justice, due process and the place of law

Clerkenwell as establishing a trend for new and more complex types of facilities within county hall. In a number of instances these went well beyond what was necessary for the efficient administration of justice. The Old Bailey (1770–4) even had a kitchen and wine vault, a covered colonnade for carriages and drawing room for the swordbearer and judge's clerk and the courts. At the Westminster courts constructed in 1823 there was a law library for the sole use of lawyers, consultation rooms and retiring rooms for lawyers. Alfred Waterhouse's Manchester Assize courts (1859–65) set even higher standards of the era for facilities with its several dining rooms for different users and rooms for barristers' clerks and consultations (Mulcahy 2008).

Segregated circulation routes

The practice of increasing segmentation of space within the courthouse is also apparent when one turns to look at circulation routes. For over three hundred years architects have been seizing on the opportunities offered by the increasingly partitioned space within courthouses to develop ever more segregated passageways between facilities. Amongst the earliest examples of segregation are the subterranean passages which began to link prisoners' cells to the defendants' area at the bar in criminal courts which emerged from the seventeenth century onwards. At Northampton Sessions House (1675–8) a vault and trapdoor for prisoners leading up into the Crown court were mentioned in 1690. Later examples of the subterraneous passage system include Smirke's Carlisle Assize court (1808–12) and Harrison's refurbishment of Chester Castle (1788–1822). It is likely that these arrangements were imposed on prisoners but soon other participants in the trial also began to demand separate routes to the courtroom. An early example of this practice can be seen at John Carr's York Assize courts (1772–7) in which judges and barristers were able to circumnavigate the central hall used by the general public by taking a private corridor which led directly to the court.

For many nineteenth-century architects the answer to the logistical problem of increasingly complex spatial practices in the courthouse came in the form of what became known as 'the concentric ring' circulation scheme. Elme's winning design for the Liverpool Assize courts competition in the early nineteenth century first experimented with the idea of surrounding central courts with a corridor from which offices and other facilities could emanate on the outside of the ring. But it was Alfred Waterhouse, with his genius for internal planning who most successfully employed the concentric ring scheme as a way of organising the movement of bodies. When judged by the standards of the professionals using the courts it is his design for Manchester Assize courts (1859–65) which proved to be the most successful of the century. During an era in which the plans of his contemporaries were often criticised for sacrificing symmetry, protocol and convenience to aesthetics (Brownlee 1984), Waterhouse's concern with the functional efficiency of the courthouse meant that he created buildings which were celebrated by regular court users as well as by architectural connoisseurs.

The particular success of his courts at Manchester was the way in which they satisfied demand for segregated routes by placing a ring around the courts whilst also creating a back and front stage to the building. The courts were situated at the centre of his design surrounded by a corridor which gave access to offices for the payment of fees, lodging of papers and other facilities. In the 'private' back half the corridors led to consultation rooms, an under sheriffs' office, a room for barristers' clerks, the bar library and robing room, rooms for attorneys, lavatories for lawyers, and the indictment office. At the 'public' fore of the ring there were separate rooms for prosecution witnesses, defence witnesses, and female witnesses and lavatories for public use. A staircase from the junction of the backstage area and the entrance to the judges' lodgings allowed barristers and judges to descend to the segregated refreshment rooms without even having to enter a public arena. Particularly worthy of note are the highly contained facilities for the judiciary, who could come from the adjoining judges' lodgings straight into the backstage area. Moreover, the judges' retiring rooms placed in the middle of the concentric ring and backing onto the courts created a second private zone for the judiciary in the midst of the courthouse. What is perhaps most ingenious is the subtlety of the zoning. Although the ring of corridors was ostensibly a public space, there was, in reality, no need for members of the public appearing as witnesses or members of the press to wander any further than the entrances to the courts themselves (Mulcahy 2008). It would appear that this ordering of facilities in terms of the needs of insiders and outsiders was fine tuned in the course of design and construction. So for instance, the original plan placed the jury retiring room in a corridor behind the court alongside a clerks' office and a barristers' robing room but in the final plan this was moved 'frontstage' and put in a central location in the public area.

Waterhouse's Assize courts were so well received that they were specifically commended to those invited to enter the architectural competition for the Royal Courts of Justice and although a relative newcomer in the architectural arena Waterhouse was appointed to advise those overseeing the construction of the national courts.[8] Moreover, although his own competition entry for the courts was unsuccessful it has been asserted that the eventual organisation of the interior owed much to his plans. George Edmund Street, who did win the commission, found that his internal design had to be much altered between the close of the competition and commencement of building works and many of the internal features he eventually adopted can be traced to Waterhouse's alternative circulation model (Brownlee 1984). Although Street's original plans allowed the general public access to the voluminous central hall this was soon corrected as a mistake and the public were provided with separate circulation routes which led directly from the street to the back of the courts (Brownlee 1984).

Despite the success of Waterhouse's circulation model, and its subsequent use in a number of modern courts,[9] it is the linear model of segregated circulation which has endured as the most efficient way to move bodies round a succession of parallel private and public zones although the imperative to create a backstage and frontstage was shared with the concentric ring system. John Soane's much debated but unexecuted early nineteenth century neo-classical designs for the new courts at Westminster provided an early prototype for this approach.[10] It is clear from contemporary drawings that, like the forbidden palace, access to courthouse facilities became more and more restricted the deeper one progressed into the building. The plan reflects the new desire to create consciously hierarchical networks of space in which the public only have access to the front segment of the building. Despite disappointment with the building as constructed, Soane was successful in creating a courthouse in which access to the accommodation became increasingly restricted the further one ventured. Westminster Hall, now free of the courts previously housed within it, was used as a *salle de pas perdus* for the other facilities. From the central hall entry could be gained directly to a narrow corridor from which the seven main courts could be accessed. Beyond this a second, rather convoluted corridor ran behind the courts and allowed judges and court officers to gain access to the offices which lay beyond. Although limited by the site, Brownlee (1984) has concluded that Soane's design imposed a new rational order on court design based on three increasingly secluded zones. The underlying logic of this plan has influenced courthouse design to the present day. The *Court Standards and Design Guide* (Her Majesty's Courts Service 2010) argues that a linear design in which the courts are sandwiched between a public zone and private zone in a long thin building is now considered superior to other designs such as those in which all other facilities encircle the courts. The layout of the newly constructed Manchester civil justice centre provides the most contemporary application of this model. In this multistorey building there are three self-contained linear 'fingers' or sections which include one which is restricted, a middle section which contains the courtrooms and a third full-height glass atrium which contains a circulation concourse and lifts for public use.

The quest for private zones within the court was not just a Victorian fetish. Anxiety about the need for dedicated space for certain participants within these 'public' buildings continued into the twentieth century. In their submission to the Beeching Commission (1971) the Lord Chancellor's office detailed that despite the £5 million per year spent on the fabric of courts,[11] judges and all the major professional legal groups were concerned about the standard of a significant proportion of the Assize estate.[12] The Commission's report legitimised concerns about the lack of segregation. They were particularly troubled by the 'common practice' of accused, litigants, witnesses, jurors, police officers and even solicitors and counsel having to jostle together in 'embarrassing proximity' (p.47) in halls and corridors which were often stacked with paraphernalia associated with other uses of the building.

An analysis of the various design guides which emerged in the wake[13] of the Beeching Report (1971) reinforce the sense of an ongoing and progressive movement to even more specialised accommodation. Within a courthouse there are now at least eight groups of people each requiring a discrete range of accommodation.[14] The staff zone alone includes 21 designated areas including offices for clerks and principals, an incident control room, store, archive, strong room and post room. Segmentation of participants in the trial has recently been further developed by new technologies which allows the evidence of a witness situated in a remote location to be brought into the court by video link and live proceedings to be broadcast into the homes of anyone who cares to log on to the internet. One of the earliest uses of video link was to reduce the security risks involved in transporting high risk prisoners from the Maze prison in Northern Ireland to the court complex to give testimony. Since then video link has been used to limit the stress to vulnerable witnesses who fear being in the same room as a person they have accused by providing a secure facility outside of the court or within a secure area in its environs.

What motivates segmentation and segregation of participants?

It is clear that changes in the spatial practices of the trial have been considerable over time. This is hardly of interest unless the reasons for change are interrogated. Movement towards segmentation and segregation in courthouse design, and the increasing popularity of division can be understood in a number of different and often conflicting ways but two particularly compelling themes emerge. The first of these is that changes in courthouse layout have been largely dictated by the changing jurisprudence of the trial. The second is that they have been motivated by fear.

In one sense the segmentation and segregation of participants was a pragmatic reaction to the increasing complexity of law and an expanding population. The seven new Assize courts of the late nineteenth century identified by Chalklin (1998) as responsible for setting trends for ever more differentiated space in the vicinity of the court all included record rooms and space for clerks. The growth in office accommodation, accounts offices and document rooms reflects the fact that arrangements whereby a local attorney would keep important documents in his office or a trunk were proving inadequate. Increased reliance on documentation, the acceptance of new types of evidence, the specialisation of law and its increasing use meant that new types of working and storage space were called for.

New concepts of due process also began to emerge in the eighteenth century which impacted on attitudes towards certain participants in the trial. The positioning of the jury in the trial is a particularly good example of the ways in which the alterations of spatial configurations of the courtroom were motivated by the changing jurisprudence of the trial. The fact that the petty jury had once shared space with spectators reflects the fact that they attended court as representatives of the local community, had powers to gather evidence outside of the courtroom and test local opinion. In time their connection with the public during the course of a trial came to be seen as a hazard and the jury came to be celebrated for its *lack* of

52 Legal Architecture: Justice, due process and the place of law

connection with spectators (Pole 2002). The development of the adversarial trial from the latter part of the sixteenth century which placed counsel at the centre of proceedings (Langbein 2003) also came to be reflected in the layout of the courtroom. As lawyers began to dominate civil trials and trials for treason, misdemeanours and then felonies and the role of the judge was reduced to that of umpire, barristers and solicitors began to dominate the well of the court from where they orchestrated the cases for the prosecution and defence.

One effect of the proliferation of private zones with dedicated circulation routes linking them to the courtroom is the impact it had on the spectacle of the trial. By keeping participants apart from each other their entry into the adjudicatory arena is endowed with much more importance. When participants no longer meet within the environs of the court but only on its stage the drama of the occasion is undoubtedly heightened by selective revelation of key figures. Having congregated in the courtroom suspense is also engendered by the judge who enters last of all and without whom the spectacle can not proceed. These theatrical devices are particularly well suited to the adversarial trial. In his work on the spectacle of the scaffold Foucault (2008) acknowledges that this particular turn in the history of English legal procedure ensured that at the end of the eighteenth century and beginning of the nineteenth century the spectacle of justice, otherwise lost when public executions were carried out behind the walls of the prison, remained. In her study of how the architecture of early American courthouses reflected the needs of increasingly professionalised lawyers McNamara (2004) has suggested that containing the spectacle of the trial within the courtroom was particularly important to advocates working within the adversarial tradition for whom the courtroom constituted their primary arena.

In the early modern period the trial was relatively unstructured and as far as criminal proceedings were concerned focused on presentation of the evidence against an accused and allowing them the opportunity to make a personal defence. The sixteenth and seventeenth century criminal trial functioned without police or public prosecutors and largely without counsel. Ordinary citizens investigated and brought their own case, leading Langbein (2003) and others to characterise trials of this era as 'altercations' between defendant and accuser. This may well account for the fact that in the courtroom at the Old Bailey built shortly after the Great Fire the defendant and witness were placed opposite each other (Gale 2006). The credibility of such approaches was seriously questioned during a number of celebrated treason trials of the late seventeenth century in which the defendants faced the State in uneven contests and perjured evidence resulted in the conviction and execution of a number of innocent people. The Treason Trials Act 1696 introduced a raft of reforms which included measures allowing defendants in trials for treason to have trial and pretrial counsel, a right which was given to defendants in misdemeanour trials and much later to those accused of felonies. In the eighteenth century the judges also began to develop the modern law of criminal evidence in order to safeguard against unfair convictions

Segmentation and segregation   53

(Langbein 2003). The corroboration, character and confession rules emerged in the criminal trial and the hearsay rule in criminal and civil proceedings. Others have explored the detail and implications of these developments in much more detail than it is possible to undertake in a book on legal architecture. For present purposes it is sufficient to stress that new approaches to what constituted sound evidence led to the institutionalisation of suspicion in the modern trial which justified segregation in the courtroom.

The importance of suspicion as a determinant of place is particularly relevant in relation to the witness. One of the purposes of admitting counsel in felony trials in the 1730s was to test the evidence of witnesses. A perception in political circles that insufficient numbers of serious property crimes were being prosecuted led to a number of government initiatives for the instigation of proceedings which brought incentives for false witnessing in their wake. The falsehood of witnesses has long been of concern to jurisprudes across cultures and is reflected in a variety of rituals associated with giving evidence such as oathtaking or ordeal and punishment of those believed to have given false testimony. But while there are occasional references to perjury as a public offence in pre-Tudor England it has been argued that there was a strong belief that perjurors should be punished by their God rather than the court. There is some evidence that a pre-Tudor verdict might be set aside because of false testimony but the view of leading authorities has been that our ancestors purged themselves with impunity on earth prior to the Perjury Statute 1563 (Underwood 1993).

As evidential practice was developed into an increasingly rational system of rules by common judges from the seventeenth century onwards,[15] concern was directed towards ensuring that evidence givers were motivated to tell the truth. Emerging concepts of 'credibility' and 'weight' meant that attempts were made to keep evidence as pure as possible. Fears that witnesses might be intimidated or extorted[16] not to testify by supporters of those they spoke against or might be influenced by the accounts of others created new incentives to keep them separate from others in the trial. The emergence of witness waiting rooms in the nineteenth century and the emerging practice of placing the witness box near to the judge are spatial recognitions of these changing attitudes, although Rock's (1993) study of Wood Green Crown court suggests that these practices were far from common by the latter part of the twentieth century. Informal practices developed by the bar and aimed at preventing the 'coaching' of witnesses also meant that rules of etiquette were established which encouraged spatial distancing of barrister and witness. Today, the concept of 'special measures' allows temporary barriers in the court to protect a witness from being seen from other participants and voice distortion can also be used for the same purpose. Witness support offices in the courthouse also signal the shift towards thinking of witnesses as vulnerable court users in need of protection. New technology also allows for the pre-recording of evidence or the virtual appearance of a witness from a remote video suite with the result that they never even have to enter the courtroom. In cases such as those involving organised crime where intimidation or retaliation against witnesses is

54 Legal Architecture: Justice, due process and the place of law

likely, segmentation continues outside the environs of the court when witnesses are placed in protection schemes.

The major alterations in courtroom design which occurred from the late eighteenth century onwards can also be seen as a reflection of shifting ideas about concepts of the public and private sphere. Changing concepts of privacy altered dramatically over this period and have been seen as constituting one of the key narratives of modernity and capitalism. The new towns and cities which emerged in the eighteenth century created new spaces in which strangers met a broad range of ranks in the newly plotted parks, coffee houses and theatres. Modernity tore apart the connection between social interaction and social networking as business in the public sphere was increasingly conducted between people without previous connections in de-personalised, even anonymous monetary transaction.

Industrialisation and political discontent in this period also rendered the public sphere ever more volatile and it was the cities and large towns in which the Assizes were held which were the main terrain of political struggle. The 'otherness' of strangers could not fail to arouse anxiety in the growing metropolises of the time and it has been argued that an initial belief in the possibility of controlling and shaping public space soon eroded. In its stead more emphasis was placed on protection from the public sphere (Sennett 1974). The idea of retreat from communal space is well demonstrated in courthouse design. The groundbreaking designs of the seven Assize courts built during 1768–71 described by Chalklin (1998) included a number of new facilities to which the elite could retire when not in court. These included separate parlours for the grand jury, magistrates and judges. Dining rooms and robing rooms dedicated to barristers and judges also became common in the years that followed as did libraries for the bar. Waterhouse's Manchester Assize courts even provided separate urinals for barristers so that they would not have to share with attorneys. These facilities reflected a desire on the part of the elite to have spaces to which they could withdraw from the hurlyburly of the public spaces of the courtroom and *sale de pas perdu.* In the courts at Maidstone (1825–6) and Lincoln (1822–8), Robert Smirke reinforced the principle of providing private space for the legal elite by allocating the area which would have normally been used as a central public hall to lawyers and judges and requiring the public to use an external arcade to access the courts from the back. These arrangements had obviously proven effective in other courts with which Smirke was involved (Smirke 1824). The fact that 'insiders' were increasingly provided with segregated circulation routes in order to circumnavigate the communal public areas completely as they moved about the building provides a strong sense of the courthouse being transformed from a communal public space to a series of private zones.

For those who feared the implications of the increase in population and concentration of people in the cities more deliberate mechanisms for detection, monitoring and control were to prove extremely attractive. Whilst many of the changes to court design discussed in this chapter can be seen as being prompted by the increasing bureaucratisation of law or developments in evidential theory, the fervour with which screens and barriers were introduced to courts is also suggestive of a process in which those with control of how space was configured began to place ever increasing stress on ordered, bounded and mappable space which allowed unpredictability to be mastered. It is certainly the case that architecture became consciously political at the end of the eighteenth century when the role of design in social engineering began to receive serious consideration by architects and politicians alike (Foucault 1984). In offering their services to the State in this way architecture has been likened to a laboratory of power in which the inconvenience of large assemblies of people could be avoided or managed (Foucault 1977). Architecture had a particularly important role to play in transforming space into place as the building, and not just the traditional signs of rank and privilege of those who used it, became a tool in the process of differentiation. Courthouses of this era established the building as a symbol of order rather than just a space where certain events occurred.

The potential for design to be used to facilitate surveillance of those who were 'out' of 'place' also came to be fully realised during this period and the physical divisions inscribed into bricks and mortar began to achieve the distancing of the other. As Foucault (1977) has demonstrated in his work on the prison and hospital, many of the disciplinary mechanisms of modernity depended on the detailed manipulation of space. Political anatomy began to work spatially during modernity through the distribution of bodies by enclosure, partitioning and division. Support for this thesis can be gleaned from the changes in courthouse design detailed in this chapter as the processes of segmentation and segregation did not treat participants in the trial as separate but equal. Instead they reflected a clear sense of hierarchy in which segmentation for the elite signalled retreat and segmentation for others reflected new forms of containment. Robert Smirke's courts at Maidstone (1824–7) provide an excellent and relatively early exemplar of this trend. The rear of each of the two courts in this sole purpose courthouse were divided into rows of seating for the public, jurymen and witnesses whilst a semi-circular inner sanctum radiates around the judicial dais and accommodates the jury and counsel.[17] Although there is accommodation for an impressive 350 spectators the plans illustrate the taste for emphasising the demarcation of the populace from lawyers and adjudicators in increasingly brutal ways. A cross section of the plans shows that high iron railings were inserted to separate the public arena from that of the inner court.[18]

The practice of providing increasingly elaborate and segregated circulation routes within courthouses of the era also tended to restrict the use of the great entrance halls of courthouses of the time to the elite. This was a practice which was followed in many of the symbolic courthouses of the late eighteenth century such as the Assize courts at York (1772–7), and continued into the nineteenth century at Maidstone (1825–6) and Lincoln (1822–8), Manchester Assize courts (1859–65), the Royal Courts of Justice (1874–82) and the Victoria Law Courts in Birmingham (1887–91). In each case separate circulation routes led members of

56   Legal Architecture: Justice, due process and the place of law

the public directly from the street to the public gallery. Reflecting on the transformation of the notion of the public sphere Sennett (1974) has argued that such practices reflect an erasure of 'alive' public space; a shift away from the idea of public space as a place where people gather to space in which people travel. He argues that this results in public space becoming a derivative of motion. Looking at such practices in the context of total institutions such as prisons and hospitals Foucault has also drawn our attention to the ways in which architecture is complicit in the undermining of an active public sphere by creating docile bodies through segregation and the canalisation of circulation (Foucault 1984). The suggestion that division prevents the emergence of solidarity or community of purpose is a compelling one. Segmentation of the courthouse makes it much less readable by occasional users and renders the temple of justice a largely secret place.

Conclusion

This chapter has looked at the internal configurations of the courtroom and courthouse and how particular participants in the trial have been placed in space over time. The history of courtroom design presented demonstrates that we have moved from the concept of an outer and inner zone in public trials to modern-day practices in which the interior of the courtroom and its environs employ highly sophisticated methods to distinguish between a variety of categories of people. The shifts in thinking about courthouse design charted in this chapter can be accounted for in a number of ways including new expectations about comfort and privacy as well as changing notions of who needs to be protected in the trial. The contemporary notion of the 'vulnerable witness' or the expectation that youth offenders or children embroiled in custody disputes should not be subjected to the challenges of the modern-day trial are now, quite rightly, commonplace. But behind these particular shifts in thinking which reflect humanitarian ideals the spatial practices in the courthouse described also provide us with excellent examples of disciplinary techniques which create a binary divide between users of authoritarian institutions. What they reveal is an ongoing fear of the public as volatile and a need to stage manage the spectacle of the trial in a way which contains emotion, noise and movement.

The creation of a back stage and front stage to the law court has served to undermine the communal nature of this public space made so obvious in early judicial gatherings. The practice brings with it the danger that the building will be incomprehensible to those who do not use it on a regular basis. One concern is that it has imposed a 'canalisation' of circulation and conscious, sometimes brutal, coding of the reciprocal relations of users of the courthouse which is inappropriate in a modern democracy. In the chapters which follow I attempt to demonstrate that in contrast to historical accounts of the trial as progressively more humane or respectful of human rights that spatial practices in the courthouse continue to be rooted in a discourse of difference based on gender and class. Not only does this

Segmentation and segregation 57

continue to render the modern courtroom a contested space but it encourages us to recognise the possibility that excessive boundary maintenance not only reflects distinctions in the respect paid to different users of the courthouse but also creates new types of cultural identities. Viewed in this way the spatial practices of segmentation and segregated circulation routes and centralised control could be seen to inscribe difference on the psyche as identity becomes shaped by the processes of marginalisation.

Notes

1 Elsewhere, these are described as the Whaddon Manuscripts (see for instance Graham 2003) or the Selby Lowndes manuscripts (*Country Life* 1922). The manuscripts came into the possession of Mr Selbey-Lowndes whose mansion was named Whaddon Hall and Corner (1863) suggests that they were either inherited by him down the female line from William Fleetwood the sometime Recorder of London or that they may have been the property of the antiquarian Browne Willis as his residence was also Whaddon Hall. It is thought that the illustrations formed part of a treatise on law as one of the illustrations has a table of contents below it. However, the four illustrations which are likely to have been drawn at the Abbey of Edmundsbury in Suffolk are all that remain of the treatise (Corner 1863). The manuscripts are currently held in Inner Temple library and are in an excellent state of repair.

2 Blatcher (1978) suggests that when sitting in Westminster Hall the Chief Justice will have shared the southern part of the hall and on occasion the marble judgment seat or throne with the Chancellor as well.

3 Bellott (1922) contends that the area behind the inner bar signified by the high back of the clerk's bench is an area for the litigants, sarjeants, utter barristers and attorneys, gaolors, prisoners and the general public. My interpretation of the illustration is that no members of the public are represented in the picture and this seems to accord with Corner's (1863) assessment.

4 For a fuller account of the work of this court see Bell (1953).

5 It is from the larger room or 'town hall', that most seats of government derived their name.

6 This practice continued in Gloucester until 1814, Leicester Castle until 1821 and Winchester until 1874. The county halls built at Derby (1657–9) and York (1673) both had one big hall which was used for two Assize courts although in both cases separate grand doorways into the two ends of the hall allowed both judges to make a ceremonial entrance to the building at the same time.

7 I was informed of this on a tour of the court which is now open as a museum.

8 This appointment was short lived as Waterhouse resigned when he decided that he also wanted to be a competitor.

9 See for instance Leeds Crown Court and the magistrates' courts built in Slough in the 1950s (*The Builder* 1957).

10 These elements of the design had to be sacrificed in order to satisfy treasury and parliamentarians' demands about style and cost. Much of the dissatisfaction with the courts which was voiced related to the lack of these facilities.

11 Half of this sum was devoted to the Assizes and Quarter Sessions. The Lord Chancellor's Office asserted that there were substantial differences between Assize, Quarter Sessions and Magistrates' courts. Because of their higher status Assize courts were generally planned to a better standard with more commodious accommodation. They estimated that the average cost of a full size courtroom in an Assize court was £250,000 in contrast to £120,000 for a Quarter Session court.

12 See for instance the memoranda submitted by Mr Justice Cantley, Mr Justice Cusack, Mr Justice Karminski, Mr Justice Waller. Those giving evidence drew particular attention to the need for increased segregation in the vicinity of the court. The majority of those giving evidence about accommodation were concerned about the lack of private consultation facilities for the parties and their advisers. Mr Justice Cantley suggested that despite the fact that designers tended to think of them as a waste of space the lack of consultation rooms was a gross error from the practitioner's point of view given their importance in conveying confidential information and facilitating settlement.

13 The first attempt at defining the core requirements of a modern court can probably be traced to the Home Office memorandum, 'The planning of court buildings' issued just before the Beeching Commission but most of the debate about centralised guidance has occurred since Beeching reported.

14 These are the Judiciary; Jury; Public; Custody; Barristers and solicitor; Catering; Courthouse Staff; and building services.

15 Some of the modern law of evidence can be traced back to the Middle Ages but most modern commentators see the seventeenth century as the period in which judicial attention began increasingly to be directed to the development of modern concepts.

16 In the United Kingdom, witness intimidation is covered by Criminal Justice and Public Order Act 1994: section 51. Under this Act it is an offence to perform an act which is intended to and does intimidate a person who the offender knows or believes to be involved with a criminal case with the intention of disturbing the proceedings.

17 It is possible that the creation of this curved 'inner sanctum' draws on precedents for semi-circular courts established by John Carr at York Assize Court (1773–7) and Thomas Harrison's Lancaster Shire Hall (1788–1821). In some locations this design was popular because it meant that the court could more easily be adjusted for use as a council chamber.

18 A modern-day example of the ways in which surveillance is provided in the courtroom is indicated by a section of the *Court Standards and Design Guide* (Her Majesty's Courts Service 2010) which deals with sightlines in relation to the jury. It requires: 'The judge must have a clear and equal view of each juror, and as far as possible be able to see what he or she is doing – e.g. reading documentary evidence or writing notes. Ideally the work surfaces in the jury box should be visible to the judge, but if, in agreement with the Court Service, upstands are installed in front of the work surfaces, the top edges of those upstands should be visible to the judge along their whole length.' (p.176.)

Chapter 4

Presumed innocent?

Introduction

This chapter and the next focus on the impact of the segmentation and segregation practices discussed in the last chapter on key participants in the trial. Whilst all those involved in the trial have been affected by the changes in spatial practices which occurred from the late eighteenth century onwards in courthouses and courtrooms, they were affected in very different ways. Changes to the courthouse and courtroom which emerged over that period provided some users with space in which they could retreat from the public areas of the court, whilst for others the impact of segregation and segmentation meant that their movement was increasingly constrained and restricted in public areas. The shifts in thinking about the spatial configurations of the courthouse which these design trends reflect are far from being politically neutral. Indeed any treatise on court architecture can just as easily focus on the courtroom as a contested space as it could on the more familiar discourse of stability, tradition and gravitas which generally surround discussions of courts. Spatial arrangements determine the mode and range of verbal interaction and emphasise the relative status of the people present and their territorial rights. But is it appropriate that historical precedents developed in different eras and reflecting different conceptions of due process continue to influence court design and render the courtroom a frozen site of nostalgia (Graham 2004)? I argue here that spacing and placing of people in the trial is strategic to their ability to participate effectively in proceedings and that insufficient account of users' experience of segmentation is paid by contemporary policy makers in the current *Court Standards and Design Guide* (Her Majesty's Courts Service 2010).

This chapter concentrates on the spatial capital allocated, claimed and enjoyed by two key figures in the trial; defendants and their lawyers. Though their enjoyment of space takes very different turns their stories are interwoven for as lawyers, and barristers in particular, begin to claim more space for themselves in the well of the court the defendant became increasingly isolated at the bar and progressively encased within a fortified dock. The importance of these accounts of space is that they make clear the ways in which two key ideals to which the criminal justice system has long aspired are undermined by architecture and design; the

60 Legal Architecture: Justice, due process and the place of law

presumption of innocence and the right to consult counsel. In the sections which follow I consider how it is that the defendant and their advisers became separated in the trial so that counsel now sit with their back to their client. How is it that the defendant, while still presumed innocent, came to be contained in a dock? These practices are particularly interesting given that in the United States, arguably the most security-conscious country in the world, the defendant and their counsel sit shoulder to shoulder in the inner area of the court. I will argue that the positioning of the defendant and lawyer in the English trial is a matter which is ripe for reform if key ideals of the criminal justice system are to be realised and enjoyed.

Differentiation

The practices of segmentation and segregation outlined in the last chapter led to the containment of people in specific areas within the courtroom and to clear visual distinctions being made between certain types of participants in the trial. One particularly noteworthy separation is that of the defendant and their advisers in the criminal trial. Whilst it could be claimed that most participants who occupy a segmented zone in the modern trial perform a separate function such as spectating, adjudicating, reporting and recording the rationale behind the separation of the defendant and their client, both of whom ostensibly perform the role of defending, is less obvious. It was not always the case that the two were separated. The fifteenth-century Whaddon illuminations clearly show sergeants at law[1] standing shoulder to shoulder with their client facing the judge at the inner bar of the court. The positioning of the sergeants suggests that even high ranking lawyers of the time were not generally granted the privilege of being allowed within the bar since from the fourteenth century until at least the seventeenth century sergeants were the highest order of counsel and the class of practitioner from which the bench were drawn.[2]

The position was slightly more complex in the High courts at Westminster where the monarch retained counsel and attorneys on a permanent basis to deal with their interests. So, for instance, the office of the Attorney General originated in 1315 when the Crown began to prosecute its own business in the Court of Common Pleas and the office of Solicitor General was established in 1461 (Attorney General's Office 2010). Graham (2004) suggests that these officers are likely to have been given positions at the central table within the bar in recognition of their status. Moreover, in his description of the figures in the Whaddon manuscripts Corner (1863) identifies one of the figures sitting below the judges within the bar of the Court of King's Bench as being the King's Attorney. A further disruption of the system whereby litigants and their counsel stood side by side occurred with the appointment of the first Queen's Counsel in 1594. Elizabeth I conferred special favour on Francis Bacon when she appointed him her learned counsel extraordinary without patent or fee and in doing so called him within the bar in recognition of the honour conferred on him (Finn 1996). From

Presumed innocent? 61

the seventeenth century onwards appointments to Queen's or King's Counsel proliferated with the result that it became more common for lawyers to sit within the bar.[3] Graham (2004) argues that by the eighteenth century distinctions about which lawyers were outwith or within the bar had begun to blur. Whether they had taken silk or not there is evidence of barristers and even attorneys sharing the central table with other officials so that this hub became the place where business between those with knowledge of the law was conducted. The layout of Warwick Shire Hall (1753–8) demonstrates that by the time that permanent fittings in courts became common it was often expected that lawyers be allowed to sit around the central table of the court, while prisoners, witnesses and jurors continue to be placed beyond it and outside of the bar. Significantly, at Warwick the architect's drawings refer to the familiar central table previously dominated by clerks in the courtroom as the 'counsel table'.

In addition to the status afforded those invited into the inner sanctum of the court, it seems likely that barristers yearned for separation from both attorneys and their clients.[4] Others have given much more substantive accounts of the professionalisation of the trial in the nineteenth century than I am able to do here[5] but for present purposes it is enough to stress that for many sociologists of the profession the key purpose of the 'professional project' of this period was differentiation (Larson 1979). According to this thesis, lawyers strove to emphasise both how they were distinct from their clients and other practitioners. Writers in the field have argued that they achieved both by placing emphasis on the technicality of law and the training necessary to understand it.[6] They also used social class and status to reinforce their claim to difference (Macdonald 1995). It is certainly the case that the eighteenth century saw the emergence of an upper and lower branch of the legal profession and increasing attempts to distinguish between them (Prest 1987). In the mid seventeenth century barristers were still being approached directly by clients but this began to be treated as unnecessary and even inappropriate and clients were increasingly expected to approach barristers through attorneys, who thus adopted the role of go-between. Barristers could also claim to be part of the regular clique in court. Members of the bar who travelled the circuits with the judiciary and other officials of the court formed a small band who became well known to each other and shared experience undoubtedly encouraged a camaraderie between bar and bench which local attorneys did not enjoy (Graham 2004).

Ethical guidelines and codes of conduct for the bar and solicitors are a relatively recent phenomena[7] but informal standards of etiquette have long governed associations between lawyers and others. Nicolson and Webb (2000) suggest that the formation in the eighteenth century of both local and national law societies in England and Wales was motivated in part by desires to establish standards and to police them but it is also the case that throughout the Victorian era the legal profession was perceived to consist of a range of different groups who shared particular interests. To most members of the bar the circuit mess rather than the Inns of Court appeared to be the most important source of

62 Legal Architecture: Justice, due process and the place of law

information about etiquette but the fact that there were numerous circuits meant that varied standards of conduct prevailed. Despite this divergence a number of key ideas about contact dominated debate and transformed approaches to space in the Victorian era. Barristers began to be discouraged from associating with attorneys and solicitors for fear that they should be seen as touting for business and members of the bar might be fined for inviting a member of the junior profession to dinner. Moreover, many messes did not allow barristers to use public transport to get to an Assize town or to stay in private lodgings for fear that they might be forced into close proximity with litigants or witnesses and accused of coaching them or altering their evidence (Lewis 1982). From a position in the seventeenth century characterised by customary lack of psychological and physical distancing in the court and its environs the position had shifted by the end of the nineteenth century to one where there were highly attuned informal rules which governed proximity of members of the bar to others within the courtroom and beyond. These changes in thinking about status and etiquette manifested themselves in a number of ways including court design and go some way to explaining why barristers moved in to take their seat at the central table while attorneys enjoyed a more ambivalent status. The distinction between the two branches also began to manifest itself in the vicinity of the court. At Manchester Assize courts barristers enjoyed separate and superior facilities to attorneys. They had their own refreshment room, library and urinals and even their clerks were given separate rooms.

The fact that barristers literally moved centre stage in the court also reflects the shift towards the adversarial trial in which lawyers played the main role in telling the defendant's story. Adversarialism is often discussed as though it was the cornerstone of common law procedure with no parallel outside of common law jurisdictions[8] but legal historians have suggested that it is actually a relatively youthful organising concept. Miller (1995) has gone so far as to suggest that '. . . an historian at some future date may look back and declare that the so-called English adversarial system was – to whatever extent it existed – a mere blip in the 900 year history of the common law . . .' (p.4).[9] It has been argued that evidentiary rules which reflected this approach only began to crystallise in the late eighteenth and early nineteenth centuries when the judiciary began to abdicate their inquisitorial discretion to lawyers in complex cases (Damaska 1997; Miller 1995). Up until the latter part of the seventeenth century most defendants were forbidden to have legal counsel and lawyers seldom appeared for the prosecution. In the era prior to the emergence of the adversarial trial it was thought that defence counsel would serve no purpose in criminal proceedings, since the defendant could speak to the court direct without interference and challenge those who accused them in person. Although lawyers were permitted to raise points of law on behalf of the accused, they were prohibited from speaking to the facts or from addressing the jury. It was argued that this 'accused speaks' model of the trial did not necessarily cause injustice to the defendant since in theory the judge was required to ensure that there was no miscarriage of justice within their court. Far fewer rules of evidence constrained enquiry and judges routinely examined the defendant and witnesses (Langbein 2003).[10] By the end of the eighteenth century important exceptions to the 'accused speaks' model of trial did exist for trials involving treason and trials for misdemeanours[11] but if the offence charged was murder, rape or arson the accused was required to make their own defence at the Assizes. As public prosecutions became more organised the notion of equality of arms became less obvious and some judges began to exercise their discretion in favour of allowing additional protection for the defendant. While the complete prohibition against counsel remained intact until the 1730s, it has been suggested that there was a noticeable trend in the eighteenth century for judges to allow the bar a limited role in felony trials (see, for instance, Langbein 2003).[12] From this point onwards it became more common for the bar to be permitted to examine and cross-examine witnesses. Despite this, it appears to be the case that active participation of lawyers in most criminal trials remained limited, even in London, until the 1780s. In Landsman's view (1980), it was only from the beginning of the nineteenth century that the courtroom contest became strikingly more adversarial and advocates and their cross-examination of witnesses were placed at the fore. The passing of the Prisoners' Counsel Act 1836, which permitted the accused to be fully represented in cases involving felonies as well as misdemeanours, is generally treated as prompting the final transformation of the criminal trial from one which focused on the accused to a fully blown adversarial contest focused on lawyers.

The shift towards a lawyer-led proceedings has been likened to a seismic shift in the history of the trial (May 2006) which it could be argued, reflected loyalty to a new and particular vision of the relationship between the citizen and State. It is no coincidence that increasing reliance on adversarial methods occurred in the same era as progressive adherence to the principles of utilitarianism and the ideal of the autonomous individual. Damaska (1986) has argued that it was in the classical period of prevailing *laissez faire* ideology that the adversarial genre of proceedings obtained quasi-constitutional backing from the assumption that cases were best seen as a dispute between two competent and autonomous parties, each of whom had their own selfish motivations for presenting the most persuasive case. According to this scheme, the role of the State was seen as a mere umpire. Its responsibility was limited to providing dispute resolution mechanisms in order to retain peace and harmony in society and to enforce the rights of individuals. Viewed in this way we can see the growth of adversarialism as the classical liberal impulse to keep the State, in the guise of the judiciary, at arm's length from the parties.[13] Damaska's (1986) thesis sits comfortably with May's (2006) argument that the Prisoners' Counsel Act forms part of the larger process of dismantling the *ancien régime* in the aftermath of The Glorious Revolution. Viewed alongside reforms such as catholic emancipation, municipal reform and the Reform Act 1832 these shifts can be seen as reflecting a distrust of a State dominated by the landed gentry and guided by self interest. In this context the right of the defendant

64 Legal Architecture: Justice, due process and the place of law

to representation was seen as a way to insulate them from abuses of power within the trial.

These heroic claims for adversarialism were not accepted without opposition. For some, criminal trials moved from a position in which the proceedings centred on the victim and accused to one in which both were effectively silenced as the professional rhetoric of law replaced the personal narratives of the parties (Schramm 2000). In the substantial debate about the role of counsel which took place in the run up to the Prisoners' Counsel Act 1836 and its aftermath commentators expressed concerns about the potential for the defendant's account of events to be manipulated by counsel and fears about the burgeoning of legal technicalities. Some of these concerns were clearly realised. The increasing complexity of legal procedure which occurred during this era meant that what was first mooted as encouraging party autonomy soon evolved into a system in which the parties were forced to find a specialist to present their case for them. Significantly, lawyers were amongst those who expressed public concerns about the increased role anticipated for advocates in the serious criminal trials. It has been argued that a considerable proportion of the rank and file of the bar opposed the introduction of the Prisoners' Counsel Act and the central role for lawyers which it anticipated. May (2006) has suggested that the opposition was sufficiently strong to be recognised in the course of parliamentary debate and is reflected in debate in the *Law Magazine*. She contends that their opposition reflected neither pure conservatism nor self interest and that some credence must be given to the bar's concern that adversarial proceedings could actually exacerbate existing inequalities in the trial. It is also possible that these newly acquired rights would not necessarily make practice at the bar more profitable, as most prisoners could not afford counsel and no system of public legal aid was available. Career-minded barristers might also have been concerned about the undermining of judicial activism anticipated by the adversarial trial.

The spatial demands of lawyers

Despite these arguments changes to the design and arrangement of the courtroom in this period suggest that claims about a lack of self interest on the part of lawyers could be viewed as somewhat disingenuous. Indeed few shifts in design better symbolise the movement from party-led to lawyer-led proceedings than those experienced by lawyers during this period. Whilst many members of the bar were vocal in their opposition to the adoption of the role of defendant's mouthpiece they were nonetheless complicit in the isolation of the defendant in the courtroom and the creation of separate working spaces for lawyer. The most noticeable manifestation of this was the reduction of the central table which had long dominated the well of the court. As this became smaller, and evolved into a small clerk's desk in front of the judicial dais, it was replaced with rows of tables and seating for lawyers. It was in John Soane's Westminster courts of the 1820s that an architect first broke away from the tradition of having a large table in the well of the court shared by lawyers and officials alike. In its stead he placed straight rows of seats for the bar and attorneys facing the judge in straight rows. Each had a moveable desk and sat with their back to the defendant. As courtroom design became more elaborate from the late eighteenth century onwards this form of planning was adopted as the norm and a typical layout of a Victorian court is shown in Figure 4.1. Whilst some have interpreted the movement of lawyers into the well of the court, where defence and prosecution often sat shoulder to shoulder, as reinforcing the camaraderie of the bar, it is a camaraderie which seems to have been attained at the expense of an equally potent symbol of the barrister's distancing from their client.

The spatial demands of lawyers[14] of this period also began to manifest itself in ways which allowed differentiation from their client in the courthouse as well as the courtroom. Accounts of barristers in the seventeenth century jostling shoulder to shoulder at the bar of Westminster Hall, clamouring for audience, taunting judges, interrupting each other and entering into brawls (Prest 1986) stand in stark contrast to the spatial demands of their successors in the nineteenth century. Indeed one of the factors which made courthouse design much more complex in the nineteenth century were the calls of lawyers for space dedicated to their use and improved standards of comfort. It is clear from analysis of the architectural plans of a number of Victorian courts, that of all the actors involved

Figure 4.1 Nottingham Guildhall (1885–8) © Ray Teece.[15]

in the trial it is barristers and solicitors, but particularly barristers, who have benefited most from changes in design. They also appear to have become more confident about asserting such claims.[16] Articles in the nineteenth-century legal press are full of regular suggestions that courts should be convenient for lawyers to use and have internal communication between courts, good ventilation, spacious corridors, waiting rooms for witnesses, a grand hall and an entrance hall, and facilities such as a dedicated bar library and refreshment room. There is evidence that some of their concerns were justified. A new court at Westminster which was added in 1857 was so badly planned that the seats for the bar were placed in the far corner of the court and meant that barristers had to scramble over and past witnesses, spectators and jurymen when they wanted to move about (Lewis 1982). The *Law Times* concluded in 1863 that, '[t]he complaint of the inadequate accommodation of our various and scattered law courts for the growing wants of the country waxes louder and louder' (p.508). A more dramatic plea was made by the same publication in relation to the impact of inadequate robing rooms on the administration of justice in the following year:

> A great cause may be lost because an eminent advocate cannot find his wig, or because a witness has got into the wrong court, or because an attorney has flurried himself in the attempt to force his way through a compact phalanx of bystanders.
>
> (*Law Times* 1864: 389)

Pleas to have solicitors and barristers involved in the planning of courts were common, but their involvement does not seem to have been the norm. Writing in 1859, the *Law Times* published a damning report on an early set of plans for the Royal Courts of Justice and concluded that they suffered from the lack of input of practising lawyers. In the words of the editor: 'As the plan stands, counsel and client may play hide-and-seek from ten till four, not to mention the positive inconvenience of having to pass from court to court through the outer air in all weathers' (*Law Times* 1859: p.138). The tension between utility and aesthetics is particularly apparent in these accounts. Those commissioning public buildings shared a desire with architects to create a façade which made a grand statement, while lawyers were often more concerned about how the interior made a statement about their status. Nowhere is this tension made more clear than in one architectural historian's reference to the Royal Courts of Justice competition as having a 'lay' panel of adjudicators despite the fact that it included the Lord Chief Justice and a recently retired Attorney General (Summerson 1970).

Limited information exists about lawyers' appraisals of courts built before 1830 as it was only after that point that specialist legal journals emerged[17] but the accolades lavished upon Waterhouse's Manchester Assize courts appear unusual in a legal press which tended to focus on the inadequacy of most courthouses. Indeed, court designs of the time seemed to suffer from the very tensions that Waterhouse was successful in managing for lawyers; the need to build a courthouse which would receive critical acclaim in architectural journals and the need for it to function well on a day-to-day basis (see further Mulcahy 2008). Ordinary lawyers were universally pleased with the building, referring to it as 'a most commodious Assize court' and 'a magnificent and most convenient building' having 'good and ample accommodation' (*Law Times* 1864a: 43).[18] One explanation for the extent of praise lavished on the building by lawyers is the fact that the amount of space dedicated to their comfort was unprecedented and set entirely new standards for courts across the country. The new courthouse provided lawyers with a generous robing room and a library with private enclaves on the ground floor. They had their own dining room on the first floor capable of seating 100 barristers, in addition to a refreshment room on the uppermost level which barristers could share with the grand jury.[19] The courthouse had private consultation rooms, rooms for barristers' clerks and attorneys, and lavatories and urinals for the sole use of lawyers. Only the comforts of the judiciary were given more prominence, and even then their facilities were contained in the adjoining judges' lodgings rather than in the body of the courthouse. As *The Derby Mercury* (1864) explained at the time, it displayed 'all the appurtenances of a great club' (p.6). Moreover, it was a great club in which barristers held a privileged status.

Spatial reconfigurations in nineteenth-century courthouses suggest that despite the bar's concerns about increasing adversarialism, differentiation from clients and others remained critical to them. However, their increasing demands for dignity and comfort, so successful when made in support of private facilities for the bar, do not seem to have been exerted in support of a dignified place for defendants in the courtroom and its environs. In fact, as we shall see, over the same period as barristers and solicitors acquired more commodious surroundings in which to conduct the adversarial encounter, the defendant was increasingly penned in within the courtroom. These accounts of the trial and its spatial context provide some justification for further interrogation of the suggestion that nineteenth-century reform was motivated by the increasing respect for the accused and their acquisition of rights to a fair trial.

Isolation

The increasing isolation of the defendant as their lawyers claimed more space in the well of the court in front of them can be seen to reflect the changing attitudes to the criminal and their punishment. Until the late eighteenth century most 'serious' offenders were punished by execution, corporal punishment, transportation or fines but by the early nineteenth century public opinion had decisively rejected indiscriminate hanging and a marked decline in those sentenced to death and the number of crimes which carried the death penalty followed. These changes were prompted by new ways of conceiving of criminal justice in which reformation became as important as retribution in penal policies and the work of reformers such as John Howard and Elizabeth Fry placed a new emphasis on the rehabilitation of

68 Legal Architecture: Justice, due process and the place of law

the individual.[20] Architects were closely implicated in the prison reform which followed in which the differentiation and classification of prisoners were seen as increasingly important if they were to be provided with an opportunity to reform.[21] Two major concepts had a fundamental impact on attitudes towards incarceration during the latter part of the eighteenth century and the nineteenth century. The first of these was the 'separate system' which required that prisoners were subdivided according to age, gender and the nature of their crime and kept in isolation. The second was the 'silent system' where prisoners slept, and sometimes spent their day, in separate cells and passed their life in total silence.[22] Evangelical campaigners focused on the increased likelihood of sinners repenting their sins if given guidance and time to reflect on their wrongdoings[23] but calls for reform often drew on a mixture of the discourses (Henriques 1972). The changes they advocated were not without challenge (see for instance, Packard 1839; Nugent 1848), but reforming groups were highly successful in having their ideas implemented.[24] The implications of this shift in attitudes for courthouse design are clear though less often discussed. At the same time as reformers were debating these issues a shift towards the isolation of the defendant in an increasingly enclosed dock are also discernible. Like those incarcerated in the new style prisons, the prisoner in the dock was removed from the temptation of association with wider society and given space to meditate on their alleged sins. The separation of the defendant from others also allowed the court to concentrate on the individual. Rather than focusing on the needs of the State, defending property or punishment the new reformist zeal claimed to look at the interests of the individual prisoner.

The appearance of the dock has varied considerably over time from simple railed enclosures to waist-height constructions with one open side and glass rooms within the court like the one shown in Figure 4.4. What is clear is that fortification of the dock first became a common feature of courts in the Victorian era. From its origins as a simple bale dock in which numerous defendants would be kept inside the courtroom or close by while awaiting trial, the dock developed into a place where an individual or co-defendants involved in one case stood alone and isolated from others. Warwick Shire Hall (1753–8) provides an example of a dock in transition between these two models. The dock consists of a large holding area at the back and a smaller enclosure at the front where an individual defendant could stand during their trial. At York (1773–7) witnesses and prisoners both stood at the bar of the court with a prisoners' holding area between them. It is clear that the need for a large bale dock which held multiple prisoners subsided as criminal procedure and evidence became more complex and the length of trials increased as a result but the prisoners' docks which emerged in their place were simple. We know, for instance, from the work of the Howard League for Penal Reform (1976) that in the 1970s it was still the case that not every criminal court had a dock. Surviving docks from the eighteenth century such as the portable dock in the Constable's room at South Molton Guildhall (1739–43) were just simple and relatively cosy wooden boxes which could easily be jumped. Maldon Moot Hall's Georgian courtroom has a step on which the prisoner would stand surrounded

Presumed innocent? 69

by a simple metal rail with four thin metal supports. Moreover, it was not always the case that such simple docks were reserved for petty cases involving low-level accusations. Drawings of the committal proceedings of the Fenian activist Patrick O'Donnell at Bow Street Magistrates' court show the dock as no more than simple iron railings to lower chest height when the defendant was sitting.[25]

Despite the many variations in practice it was in the nineteenth century that we see citadel-like constructions emerging in the centre of newly built courtrooms. Modest boxes and rails gave way to more secure wooden chest-height docks with subterranean passages and a number of these are still available for inspection in historic courthouses.[26] Docks became larger, higher and commonly dominated the court. For instance, illustrations of the Old Bailey in 1882 suggest that when the prisoner sat down in the dock only their neck and head were visible.[27] Fortification of the dock also became more common during this period. A dock at St Albans Town Hall and Courthouse dating from the era has large protruding spikes of at least a foot in height above a chest-height wooden dock.[28] A similar fortified dock in Newcastle Guildhall which dates from the seventeenth century but has some nineteenth-century fitting is shown in Figure 4.2. The typical Victorian dock effected the presentation of the defendant in different ways depending on their height and build. At one extreme children, still tried during this era in adult courts,

Figure 4.2 Dock at Newcastle Guildhall (1655–8) © Steve Milor.[29]

70 Legal Architecture: Justice, due process and the place of law

had difficulty being seen by those who judged them. Accounts of trials in the *Illustrated Police News* include several examples of defendants whose heads were only just visible over the top of the structure (see for example *Illustrated Police News* 1867a, 1867b, 1867c).

The clear differentiation of the defendant from others beyond the bar increased the dramatic effect of the defendant's entry and exit to the court and focused

Figure 4.3 Approach to the dock from underground cells in St George's Hall, Liverpool © Linda Mulcahy.

Presumed innocent? 71

attention on the danger they posed to others. The ritual of entering and leaving the dock appears to have developed a new poignancy during this period. Newspapers of the nineteenth century regularly open their accounts of a trial by reference to the defendant being 'placed' in the dock and conclude them with references to their demeanour on exiting it and being 'taken down'. Accounts of the latter range from stock phrases involving the prisoner looking blank and collapsing to displays of angry threats of revenge. The prominence of the Victorian dock in the popular imagination is further emphasised by the practice of including illustrations of the trial which focused on the defendant's demeanour in the dock, a practice which continues to this day. This emphasis on the area provides an interesting inversion of Bentham's panoptican ideal in which the few surveyed the many. Most significantly, the attention lavished on defendants in contemporary accounts of the Victorian trial served to reinforce the ways in which the process of administering criminal justice becomes a form of punishment in its own right. As Rosen (1966) has argued in a contemporary context:

> For a man (or woman) to stand in the dock is a humiliating and degrading experience. He is isolated from his legal adviser. He is a man apart. He is the cynosure of all eyes. He is placed, as it were, in a pillory, and must feel he is an object of scorn and derision. Persons who have been acquitted have stated that their sojourn in the dock has been the part of their ordeal they have found hardest to bear. (296–7).

Given the aspirations of the *Court Standards and Design Guide* (Her Majesty's Courts Service 2010) that the modern courthouse should be a more democratic space than in previous eras one might imagine that the construction of the dock, and indeed its very presence in criminal courtrooms might have been challenged in recent decades. The *Guide* does recognise the significance of good acoustic qualities in allowing the defendant to be heard across the large space of the courtroom. The importance of the defendant and judge being able to see each other is also stressed as is the facilitation of the defendant taking an 'appropriate' part in proceedings. Specific reference is also made to the importance of avoiding a 'cage like oppressive environment for the defendants' (p.109). The templates for docks prescribed by the Guide are not as visually oppressive as their Victorian counterparts and the dock in the Magistrates' court is expected to be placed along one flank of the court giving the defendant a better opportunity to move to the front section of the court. This undoubtedly undermines the feeling of isolation experienced in most modern Crown courts where it continues to be recommended that the dock is placed at the very back of the court.

Despite these expectations the *Guide* remains committed to the presence of the dock[30] with the result that many defendants continue to be marginalised in the trial. Regardless of its concerns about oppression, the *Guide* also emphasises the dangers posed by defendants in the trial rather than protection of their presumption of innocence. Section seven specifies that the standard dock in the Magistrates' court should be sited so as to be as far as possible from both the public entrance to the courtroom and the magistrates' entrance to the bench. Other references to the dock include the recommendation that an affray alarm and whisper telephone should be installed for use by the dock officer should the need arise. The defendant is expressly denied a writing shelf on which they might write instructions to their counsel, whether as part of the dock-front or seating, as it is anticipated that this might serve as a step in an escape attempt.

New materials also allow the dock to become cage-like by different means. The *Guide* specifies that the dock-front should be a minimum of 150mm thick and 1350mm high, and of robust construction. It should be surmounted with either a semi-circular top to inhibit handholds in the event of an escape attempt, or by an additional 350mm of forward-projecting toughened glazed panels or rail protections. The dock is expected to be higher where it meets the adjacent public seating (1525mm above finished floor level) in order to prevent members of the public and the defendant from seeing each other clearly. Whilst this measure protects the defendant from members of the public who might want to intimidate them it also has the potential to isolate them from their supporters. In many courts the modern dock is best understood as a room within a room, or as a transparent cell within the courtroom. The secure dock template requires that the security screen above the wooden base will be made up of a series of transparent bars with 25mm gaps between them at intervals of 45mm in order to allow the defendant to pass notes to their lawyer. Contemporary examples of this arrangement can be seen for instance at Southwark Crown court in London.

New technologies now allow us to revisit the configuration of the dock in ways not anticipated by the *Guide*. Live video link provides incarcerated defendants with opportunities to deliver their evidence from the prison in which they are being held without even having to attend the court. Moreover, Jack Straw has recently launched a virtual court pilot scheme in Kent in which the defendant 'appears' in court via video link from the police station at which they have been charged that day. Rather than improving access to justice, as has been suggested,[31] this practice would appear to revert to the much disapproved of nineteenth-century practice of courts of summary justice having a close physical association with the prosecution by means of their attachment to police stations. Such close association between the police and the trial has been seen as calling the impartiality of proceedings into question in the past. It would seem that the practice of holding courts in the same complexes as police stations has been abandoned only to be reverted to once more through the medium of information technology. The video released to advertise the new scheme in Kent suggests that when the defendant and their lawyer appear in court from a room in the police station, that room becomes a combination of dock, witness stand and consultation room, although this does allow lawyers to be reunited with their client once more. The launch video shows the duty solicitor sitting shoulder to shoulder with their client as they address the television screen which links them to the court.

Challenges to the isolation of the defendant

The distancing and isolation of the defendant from those determining their fate has not gone unchallenged. In the early nineteenth century, as the fortified dock gained in popularity, a flurry of cases were heard in which the positioning of the defendant was questioned. These centred on concerns that the defendant was being excluded from the inner sanctum of the court and separated from their lawyer. The cases make clear that the positioning of the defendant relied partly on the severity of the accusations made against them. While it was permissible for the defendant to sit behind counsel in the body of the court in cases involving misdemeanours, the presumption was that this privilege was not available for defendants accused of felonies.[32] Indeed, in *R v Lovett* (1839) it was argued that a defendant on a charge of misdemeanour might be allowed to take a place at the table of the court.[33] It was also made clear in *R v Carlile* (1831) that while an unconvicted defendant on bail ought to stand in the dock it was not usually insisted upon.[34] In the State trial of Horne Tooke a concession was granted to the defendant on the basis that he was suffering from ill health and unable to stand at the bar for long periods. However the court was keen to stress that it should be treated as an exceptional case[35] and generally the courts have been reluctant to grant such requests. In *R v Pedro de Zuleta the Younger* (1843) a request from a defendant that they be allowed to sit beside their counsel because of language barriers was turned down. The suggestion that those with high social status should be afforded the right to sit with counsel inside the bar also rarely met with the approval of the court.[36] Indeed in *R v George* (1840) the court referred to several cases in which even peers of the realm had to stand at the bar throughout their trial. It was explained that:

> On the trial by the Peers of Lord Ferrers, in 1760, for the murder of Mr. Johnson . . . and on the trial by the Peers of Lord Byron, in 1765, for the murder of Mr. Chatworth, the noble prisoner in each case, though a peer 'was brought to the bar by the deputy governor of the Tower, having the axe carried before him by the gentleman gaoler, who stood with it on the left hand of the prisoner, with the edge turned from him; the prisoner, when he approached the bar, made three reverences, and then fell on his knees at the bar'. And on the trial of Lord Kilmarnock, and other Peers, for high treason in 1746 . . . the same forms were observed. In the Duchess of Kingston's case for bigamy, tried by the Peers in 1776 . . . the Duchess was brought to the bar by the Deputy Gentleman Usher of the Black Rod . . . and 'the prisoner, when she approached the bar, made three reverences, and then fell on her knees at the bar' (p.923)

Some exceptions to this rule appeared to be permissible where royalty were concerned. It is suggested in *R v George* that in the trial of Mary Queen of Scots she may have been placed within the bar as evidenced by a contemporary drawing.

74 Legal Architecture: Justice, due process and the place of law

Moreover, though positioned at the bar Charles I was allowed to sit on a crimson velvet chair during his trial in Westminster Hall and was sufficiently close to his counsel to be able to tap him on the shoulder with a silver cane when he desired his attention.[37]

In other instances defendants claimed the right to sit at the central table when acting as their own counsel. In *R v Carlile* the defendant conducted their case without assistance and took a seat at the table within the bar at the start of proceedings. He was asked to move and instructed by Mr Justice Park that only the bar had a right to the seats concerned and that no one else was allowed to occupy them. However, in a discussion of the unreported case of *Staley v. Hunt* (1834), referred to in argument it was suggested that practice in the provinces might differ. It is made clear that Mr Hunt had been allowed to conduct his own case in the body of the court at a trial conducted at Nisi Prius Assize but this was not allowed when he appeared at the Court of Exchequer in London. Indeed, it is stated in the discussion of the case that the distinction 'within the bar' did not exist at Nisi Prius. Moreover, in *R v Vincent and Edwards* (1839) the defendant conducted their own defence at the table of the court at the Nisi Prius side of Assizes even though he was in custody. These cases suggest that there appears to have been little recognition in English jurisprudence of the impact that such isolation might have on the conduct of the defence.[38] What remains particularly curious is that as all defendants began to acquire the right to full representation in the early nineteenth century their isolation in the court became more pronounced with the gradual encasement of the defendant in the dock. It is to the impact of this use of space on the defendant that we now turn.

Degradation

There has been a steady trickle of concern expressed about the presence of the dock in the English trial and its impact on the defendant's ability to participate in their defence. As long ago as the nineteenth century a commentator on proceedings in the Old Bailey recalled the words of a well known counsel defending a 'singularly ill-favoured prisoner' when addressing the jury:

> Gentlemen, you must not allow yourselves to be carried away by any effect which the prisoner's appearance may have upon you. Remember, he is in the dock; and I will undertake to say, that if my lord were to be taken from the bench upon which he is sitting, and placed where the prisoner is now standing, you, who are accustomed to criminal trials, would find, even in his lordship's face, indications of crime which you would look for in vain in any other situation!

(Jones 1871) (p.4)

In her seminal two year study of Magistrates' court proceedings in the twentieth century Pat Carlen (1976) argues that the exploitation of courtroom space has a paralysing effect on those who are not regular users of the court system. In part this is attributed to the various ways in which conventional use of space is disrupted in the courtroom. So, for instance, accepted and familiar modes of conversational practice are perverted so that confessions and highly personal stories which would normally be told in close and intimate spaces are conducted over much longer distances and in the presence of strangers. In this way space can be seen as contributing to a ceremonial stripping of dignity. She found that the distance between the defendant and other players in the court meant that difficulties of hearing were endemic and lead to constant requests for things to be repeated or for people to speak up. Significantly, all the participants in her study complained of the sterile theatricality of the courtroom in which temporal and spatial conventions were successfully manipulated to produce a disciplined display of justice in which 'alternative performances evocative of unpermitted social worlds' (Carlen 1976: 12) were suppressed. In her words:

> Of all the main protagonists the defendant is the one who is placed farthest away from the magistrate. Between the defendant and the magistrate sit clerk, solicitors, probation officers, social workers, press reporters, police and any others deemed to be assisting the court in the discharge of its duties. Spatial arrangements ... which might signify to the onlooker a guarantee of an orderly display of justice, are too often experienced by participants as being generative of a theatrical autism with all the actors talking past each other (p.50).

In his call for its abolition Rosen (1996) has argued that the dock is a prominent and ugly structure in the middle of the court which disfigures the courtroom, mitigates against the presumption of innocence and is a mark of degradation (see also Barrister 1978).

More recently the human rights implications of these spatial practices have led to a resurgence of interest in the subject and calls for a presumption against the docks use in English courts by campaign groups and the Law Society (Howard League 1976, Law Society 1976[39]). The trial of Thompson and Venables for the murder of Jamie Bulger provides a rare contemporary example of the ways in which the position of the dock was said to undermine article six of the European Convention on Human Rights. In *V v United Kingdom* (2000) Venables was successful in asserting that the cumulative effect of several aspects of the organisation of his trial meant that he had been denied a fair hearing and that this gave rise to a breach of the Convention. These included the accusatorial nature of the trial, the fact that it was held in a public court as adult proceedings, the length of the trial, the presence of a jury of 12 adult strangers, the overwhelming presence of a hostile media and public, and the disclosure of the defendants' identity. Most significant for the purposes of the argument being pursued here is the fact that the physical layout of the courtroom was found to have contributed to the undermining of the defendant's right to a fair trial. It was determined that the raising of the dock, undertaken in the hope that it would ensure that the defendants could see what was going on, actually had the effect of increasing their sense of discomfort and exposure. It transpired in the course of evidence that the defendant had paid very little attention to what was going on around him and as a result he had been poorly placed to instruct counsel. Venables cried throughout most of the trial and claimed to have spent much of the time counting in his head or making shapes with his shoes. Expert evidence on his state of mind suggested that he had felt better after the first three days in the Crown court, but only because he had stopped listening to the proceedings.

The use of the dock in the modern trial has also been the subject of criticism in other jurisdictions. In the recent decision of 2007 the Supreme Court of Victoria at Melbourne in *Benbrika (Ruling no 12)* it was held by Bongiorno J., that a perspex screen which heightened the dock and divided it into segments separating the defendants into groups of two suggested that the defendants were people who warranted being guarded against[40] and undermined the presumption of innocence. As the judge argued such practices '. . . cut the accused off from the courtroom in such a way as to render the accuseds' presence hardly more real than if they appeared by video link' (para.28).

American case law has characterised the dock, in those states where it continues to be used, as a brand of incarceration which is analogous to the defendant having to wear prisoner's clothes during their trial (Doerksen 1989–90). In *Walker v Butterworth* (1978) the court determined that the dock was an anachronism in the modern trial which should have been abandoned years ago and that it

Figure 4.4 A modern glass dock © Her Majesty's Court Service.

was capable of undermining the presumption of innocence. Whilst the importance of being able to identify the accused in the court and the minimisation of danger of harm were important considerations in the trial, it was none the less concluded that these were achievable in less prejudicial ways than placing the defendant in a dock where they posed no danger.[41] In *Young v Callahan* (1983) it was reiterated that, like appearance in prison garb, confinement in the dock is a constant reminder of the accused's condition as a prisoner and serves to erode the presumption of innocence.

When called to account about the ongoing presence of the dock in the criminal trial by the Howard League for Penal Reform (1976) and the Law Society (1976) previous governments have remained unwilling to change their policy. Experts in court security have acknowledged that the majority of prisoners can be expected to behave well in court but anxieties about the potential risk of an outburst by the defendant, which placed court staff at risk, or an outburst by members of the public, which might endanger the defendant, have continued to prevail. Significantly, concern about allowing the defendant to sit in the well of the court next to their lawyers has not only come from those with responsibility for security within the court. Files on the debate surrounding a suggestion by the Secretary of State in the late 1960s, that reforms be piloted, appear to have met with a considerable amount of opposition by the Bar. A memorandum circulated with the Lord Chancellor's Office at the time suggested that the Bar Council were of the view that the accused must have 'their proper enclosure' and that it was absurd to suggest that the presence of the dock prejudiced the jury against the defendant (Lord Chancellor's Office, 1999).

English policy in relation to the dock is particularly interesting when compared to that being adopted by other Western jurisdictions such as the US, Holland and Denmark where the accused sits next to their lawyer as a physical manifestation of the ideal of equality of arms. Courts in the United States do not appear to have had the same attachment to the dock as their English counterparts[42] and even in states where it survived the settlers crossing it has long been challenged as interfering with the defendants' right to counsel as enshrined in the sixth amendment to the American Constitution, the requirements of due process and the presumption of innocence. Shepard (2006) argues that although the use of a dock continued in the Eastern seaboard well into the twentieth century most US courts had ceased to confine the defendant during trial by the end of the nineteenth century. The general approach has been to free the defendant from the dock and to allow them to cross the bar that separates spectators from participants in order to sit next to counsel. Indeed, American courts have interpreted the right to assistance from counsel as only being satisfied where they are nearby and can be consulted easily (Rosen 1966).[43] Shepard (2006) has drawn attention to the various ways in which the American courts have gone beyond granting the criminal defendant the right to sit within the inner bar of the court by giving an increasing amount of autonomy to the defendant to choose where they will sit once there. In a case involving the former head of the Enron Corporation an attorney argued that rather than being

78 Legal Architecture: Justice, due process and the place of law

confined to the table traditionally set aside for the defendant furthest from the jury box, he and his client should be allowed to sit at the table directly across from the witness stand in order that they could enjoy unobstructed and uncluttered face to face confrontation with the witnesses. Relying on notions of fairness and common sense the judge allowed both the prosecution and defence use of the table nearest the witness box when presenting their case. Reflecting on the case, Shepard (2006) argues that, the American courts are granting the defendant more and more freedoms, a fact which provides a stark contrast to the position in the modern English court.

Conclusion

The Whaddon illuminations demonstrate that little attention was paid to the dignity or appearance of prisoners in medieval trials. The illustration of King's Bench shows unkempt and shackled defendants standing at the bar of the court, and drawings of the Court of Exchequer show the litigants awaiting trial being held in a cage. In many ways one might expect this of a feudal society in which the emphasis was on punishment rather than reformation. The transient nature of court furniture also meant that it was not always practical to have a permanent area in place for litigants who posed a threat to security. The political and architectural context of the modern trial is significantly different and contemporary policy makers would no doubt claim to have very different approaches to the treatment of the defendant in the courtroom. Despite this there is considerable socio-legal evidence that appearing as a defendant in a criminal trial is a degrading experience and that current design standards exacerbate the inevitable stress and shame associated with arrest and trial rather than relieving it until such time as the defendant might be declared guilty.

Theorists of criminal law and procedure regularly assure us that there are certain principles which lawyers hold dear in the trial. One of these is the presumption of innocence, the other that defendants have the right to independent advice in conducting their defence. Both are undermined and indeed rendered almost possible by the current design of the English courtroom. Defendants are not always placed in the dock in criminal trials. When they are on bail or being charged for a minor offence it is not always required that they should be placed in a dock. However, although the judge has a wide discretion to alter the placing of the parties, requests for the defendant to be allowed to sit alongside them seem to be unusual. The result is that legal advisers commonly sit removed from their client in court. Even when the client is removed from the dock they are commonly directed to sit behind their counsel.

Far from facilitating contact with the defendant's legal adviser, the physical distance between dock and lawyers' seating seems to militate against it. The modern trend towards docks with a wooden base topped by glass slats further exacerbate contact. One of the purposes of placing gaps between the glass slats may be to allow the defendant to pass a note to their advisers but the bar on a writing shelf in enclosed docks and the absence of pens and paper in practice do not encourage the exercising of this right. The defendant's isolation within the dock and the expectation of passivity is further promulgated by the fact that when the defendants elect to give evidence in their defence they are moved from the dock to the witness box in the body of the court. The result is that for many defendants on trial their experience of the dock will be one of being contained in a room within a room where silence is expected. They can see what is happening but not engage effectively. They can hear what is happening, but often not as effectively as those placed in the well of the court. They enjoy a right to counsel during the trial but are discouraged from exercising it by their physical environment and those who have played a part in designing it.

Notes

1 Though the various personnel in these fifteenth century illustrations can be difficult for modern researchers to distinguish, the sergeants at law can be identified by the white linen coif they wear on their heads (Corner 1863). Many of the participants in the illuminations wear part-coloured robes. Corner (1863) suggests that the use of part-coloured gowns probably originated from the granting of liveries of chosen colours by great lords to their friends, servants and bondsmen. Sovereigns tended to disapprove of the practice and a number of Acts of Parliament were passed which limited the giving of liveries. Sergeants at law seem to have been granted an exception as they wore their gowns long after they were abandoned by others.

2 They first began to form a rival elite and then surpassed the established order of the sergeants. On being appointed these lawyers ceased to be utter barristers and withdrew from their Inn in order to become a member of one of the two Sergeants Inns, one of which was based in Fleet Street and the other in Chancery Lane. Up until 1845 they enjoyed an exclusive right of audience in the Court of Common Pleas.

3 By the nineteenth century the bench were still drawn from sergeants but the practice of appointing senior counsel to the 'order of the coif' for a brief period before appointment to the bench had become a mere formality (Finn 1996). For a more detailed account of Sergeants see Hastings (1947).

4 The General Council of the Bar was established much later in 1894 but statutory recognition of the professional status of barristers had been in place since 1532 (Raffield 2004). The interests of barristers had been overseen by the Inns of Court, which had effectively operated as Guilds since medieval times. By Tudor times a complex educational system for trainee barristers was in existence and the complex rituals involved in these reflect a clear group consciousness by the sixteenth century when the benchers of the inns acquired a monopoly over admission to the profession. Although a new breed of more competitive barristers with less respect for these group rites emerged in the wake of the Tudor litigation explosion a high degree of self-conscious exclusiveness and mystery continued to be encouraged. Certainly by the eighteenth century barristers on circuit during the Assizes had institutionalised their status as an elite band of lawyer's semi-formal messes where particular standards of decorum and etiquette were developed and adhered to.

5 But see Prest (1986), Brooks (1986), Lemmings (2000).

6 Differentiation from 'rogue' practitioners was achieved by the formalisation of entry requirements and the development of powers to discipline those who appeared inadequate. A study of law journals in the latter part of the nineteenth century suggests that there was widespread anxiety amongst attorneys about rogue practitioners in the

County courts and the need to uphold high entry standards. The imposition of a professional exam in 1836 and a second in 1862 which served to diminish the number of solicitors eligible to qualify appears to have been a reaction to such concerns (Abel 2003). Over the same time the bar appear to have became increasingly sensitive about relationships with attorneys who they clearly saw as the lower branch of the profession (Lemmings 2003).

7 Maughan's Treatise on the Laws of Attornies published in 1825 went some way to addressing expectations but it was not until 1974 when Lund's (1960) lectures into professional conduct were converted into the first edition of Law Society Guidelines on conduct. The bar published its first code of conduct as late as 1980 but the most recent version remains relatively short. Nicolson and Webb (2000) suggest that the publication of a code was necessitated by the break up of a homogenous community as women, the working classes and ethnic minorities entered the bar. Lund (1960) claimed that many of the matters covered in his lectures were those on which there had previously been no authoritative or accessible statement made.

8 See for instance The Royal Commission on Legal Services (1979) and the judgments of Lord Roskill in *Bremer Vulkan* (1980) and Lords Fraser and Scarman on appeal in 1981.

9 Whilst it is clear that all legal systems have adversaries not all have an adversarial orientation. The term adversary has a much longer heritage than the term adversarial which appeared in the *Oxford English Dictionary* in relation to judicial proceedings as recently as 1972 (Miller 1995).

10 Langbein (2003) dates the transformation towards a lawyer-led trial as occurring in the period from the 1690s to 1780s.

11 Persons accused of high treason were granted the right to representation by the Treason Trials Act 1696 which was passed in the wake of a series of miscarriages of justice caused by politically motivated trials and convictions. The parties in trials for misdemeanours had also been represented since medieval times, an apparent inconsistency which may attribute to the fact that many misdemeanours related to complex property-related 'offences'.

12 They were still not permitted to address the jury or make opening or closing speeches. Historians have also demonstrated that lawyers were also becoming more involved in preparing cases for court over this period. This is particularly the case in the sphere of civil law. As commercial transactions became more complex a number of large institutions such as the Post Office or Royal Mail employed solicitors and attorneys to prepare the charge and evidence: solicitors may not have appeared in courts as advocated but they increasingly had good reason to be in its environs and instructing the bar (May 2006).

13 Significantly, as the parties and their advisers have increased their level of control the formal role of the judge has diminished with the result that the judiciary are relatively passive in the adversarial model. They are generally dependent on evidence and witnesses produced by each of the parties. They must only rely on what they 'hear'. Whilst a judge will occasionally make their own inquiries of a witness their intervention is comparatively limited since they are in danger of putting their own neutrality at risk.

14 There is a danger of overstating the homogeneity of the legal professions. There have probably always been fault lines between the provinces and London, the type of practice undertaken and the solicitors profession and the bar as well as divisions based on ethnicity and class (see for instance McQueen 2003). But it is the case that law societies which sought to promote the collective interests of attorneys appeared from 1739 in London and the 1770s in the provinces. A national Law Society was formed in 1825 to protect the interests of attorneys, solicitors and proctors.

15 See further http://www.nottingham21.co.uk

Presumed innocent? 81

16 As well as the changes to their status which came about in the shift towards the adversarial trial the need for barristers was also evidenced by a sharp rise in the amount of litigation being conducted. The number of people being prosecuted at the Assize and Quarter Sessions rose from 4,605 in 1805 to 31,309 by 1842. Moreover, as lawyers became more involved in trials the length of proceedings increased and the need for professional representation more acute. Those accused of treason were allowed to have counsel from 1696 and by 1730 lawyers were appearing at some of the more ordinary criminal trials at the Old Bailey. By 1800 Graham (2003) reports that they were more common, but their attendance was still not routine. Moreover, it was not until 1836 that defence counsel in criminal matters were allowed to sum up in cases. The use of lawyers in civil trials has a much longer heritage.

17 *Justices of the Peace* was established in 1837, *The Law Times* in 1843, *County Court Chronicle* in 1846, the *Solicitors Journal* in 1857, and the *Law Chronicle* in 1858.

18 A comparison of the winning entry with the final plan shows that the size of the bar's library actually increased during the planning period. This occurred at the expense of one room for barristers' clerks and one consultation room for use by barristers and attorneys.

19 The bar were not strangers to the importance of defining boundaries between their brotherhood, the proletariat and the State. In his seminal work on cultures and images of law Raffield (2004) charts how the design of the Inns of Courts, and in particular their expansion in the litigation explosion of the Tudor period, was underpinned by a desire to reinforce the notions of the bar's autonomy and exclusivity. His work also emphasises the centrality of the ritual of shared dining to the bar which may in some small part explain their insistence on exclusive dining facilities in the new purpose-built courthouses.

20 A key implication of these changes was that incarceration, previously viewed as a short term measure leading up to trial or sentencing now began to be seen as an important form of punishment in its own right. The most common type of prison erected in the seventeenth and eighteenth century was the small town lock up in which disorderly individuals could be contained for short periods. Brodie *et al* (1999) assert that Newgate (1770–80) in London was the first English prison to assert a more monumental architectural presence.

21 See in particular the designs of architects such as William Blackburn and later George Byfield and Daniel Alexander.

22 This system was introduced at Pentonville (1840–2). The thinking behind the building is that of William Crawford and Joshua Jebb who argued that prisoners should be given separate cells, that they should be called by number and not by name, that total silence should be maintained, that head masks should be worn in the exercise yards and that when in church they should be separated by individual boxes. A surviving example of this system can be seen at Lincoln Castle.

23 The separate and silent system can be seen as being motivated by a number of different factors which might also be seen as providing a context in which a more prominent dock in the courtroom became possible and desirable. The high incidence of squalor and new interest in hygiene and the prevention of disease motivated the desire to segregate prisoners from each other. Rational utilitarians led by Bentham placed emphasis on the possibility of efficient systems of incarceration which would deter others from committing crime in which idle hands could be put to work and the costs of supervision reduced.

24 See further The Penitentiary Act 1799, The Gaol Act 1823.

25 See Herber (1999), p.54.

26 These include Boston Sessions House (1843), Devizes Assizes (1835), Durham Crown court (1809–11), Ely Shire Hall; the remodelled Leicester Castle (1821); Lincoln County Hall (1822–8); St George's Hall (1847–56), Nottingham Shire Hall (1769–72);

82 Legal Architecture: Justice, due process and the place of law

Presteigne Shire Hall (1826–9); Carlisle Assize courts (1808–12); York Assize court (1773–7). Photographs of all of these historic courts can be seen in SAVE (2004).

27 See the illustration in Herber (1999), pp.52, 55, 58.

28 See illustration in Herber (1999), p.103.

29 See further http://newcastlephotos.blogspot.com/

30 The *Guide* also anticipates that different sorts of docks will be needed in different types of court. The proportion of 'secure' to 'standard' docks will depend on local conditions and need. A secure dock is defined as one that is so constructed as to prevent a defendant being able to leave the dock by breaching its structure, scaling over the top or through a door into the well of the court or custody suite unless the door has been opened to allow them to do so on the instruction of the court or custody officer. The standard or non-secure dock is used where there is no particular threat of violence or escape from the defendant (see further pp.109–110 and Appendix 6.1).

31 Online. Available HTTP: <http://www.justice.gov.uk/news/newsrelease120509a.htm>

32 See *R v George* (1840) and the report of *R v Egan* (unreported) discussed in the case.

33 See Wakefield's Case.

34 In that particular case Mr Carlile was allowed to sit by the Clerk of arraigns although it was not made clear where that place in the court was.

35 *R v Horne Tooke* 25 State Trials p.6. Mr Justice Lawrence made clear: 'I have consulted my Lords the Judges who are present; they feel themselves extremely disposed to indulge you on the score of your health; they think that it is a distinction which may authorise them to do that in your case which is not done in other cases in common; they cannot lay down a rule for you which they would not lay down for any man living; but if your case is distinguishable from the cases of others, that does permit them to give you the indulgence which you now ask'.

36 See the report of *R v Douglas* in *R v Zeleta* (1843).

37 On his impeachment for high crimes and misdemeanours in 1806 Lord Melville sat on a stool within the bar and did not go on his knees but it is likely that the trial was short.

38 In *R v Douglas* (1841) the defendant was allowed to have three friends sit with him there during the course of the trial (1 Car and M 193) but this practice does not seem to have been common.

39 An undated memorandum by the Council of the Law Society entitled 'The use of the dock in criminal courts' appears as the first appendix in Howard League (1976). However, it is also referred to a decade earlier by Rosen (1966).

40 See also online. Available HTTP: <http://sheikyermami.com/2008/08/31/nsw-terrorism-trial-judge-wants-glass-screen-re>

41 See also *Commonwealth v Moore* (1979); *Bumper v Gunter* (1980).

42 Doerksen (1989–90) suggests that the dock was not an important feature of English courthouses by the seventeenth century and so there was no reason to expect that it would be exported to the US with the early settlers. He suggests that it was slowly introduced to some jurisdiction as the English began to tighten control over the colony's legal system. Shepard (2006) argues that the dock survived the English crossing but Doerksen's version is much more persuasive in the context of my own research into the state of the seventeenth-century court in Chapter 2.

43 See for instance *People v Zamora* (2000) 152 Pacific Reporter, 2nd series 180.

Chapter 5

Open justice, the dirty public and the press

Introduction

This chapter explores the extent to which segmentation and segregation of the courthouse and courtroom have been fuelled by a fear of the public and the way in which the idea of an 'open' public trial has come to be perceived of in spatial terms. Fear of uncharted or unscripted performances and physical and mental 'contamination' emerge as strong themes in the historical accounts of design presented. I argue that these practices have served to degrade the importance of the spectator in the trial and allowed them to be treated as peripheral to the administration of justice. One reaction to such concerns about how space is used to engender discomfort might be to suggest that courts are supposed to be daunting places in which participants are encouraged to reflect on the gravity of law and legal proceedings. According to this view the encounter should be extraordinary and memorable. The court is not, and arguably should not be, akin to an academic seminar. Violent outbursts from the public gallery, harassment of the jury and intimidation of witnesses are far from unheard of. My purpose is not to suggest that some surveillance and discipline in the courtroom are inappropriate *per se* but rather that design can serve to glorify and humiliate in inappropriate measure and in doing so undermine the contention that the courts are 'open' to the public in any meaningful way. The thesis I pursue here is that as the public have become increasingly contained within the courthouse the possibility of participatory justice has been seriously constrained as a result.

The lack of sustained academic commentary on treatment of the public in court can in part be explained by the fact that greater prominence is often given to the press as agents of the public who act in their interest to publicise trials. The press have fought hard for the right to attend, comment on and report proceedings and it is undoubtedly the case that they have played a significant role in uncovering miscarriages of justice and procedural error. Newspaper and television journalism has also played a part in disrupting traditional notions of the trial by bringing the drama of proceedings into the living rooms and television sets of the public where their consumption of legal narratives can not be monitored by court staff. These achievements have also brought tension in their wake as the accounts

84 Legal Architecture: Justice, due process and the place of law

of participants in the trial are mediated by journalistic 'creativity' and the profit motive of newspaper magnates. Whilst many members of the press have undoubtedly served the public interest in uncovering injustices there is also an important tension to be recognised. Just as the defendant in the modern trial could be said to have been silenced by their lawyer, so too it could be argued that the participation of the press in the trial has in turn justified the sidelining of the public they claim to represent. This chapter seeks to explore these dynamics. It argues that if the trial is to maintain its moral legitimacy greater efforts must be made to involve the public in the justice system as participants rather than mere spectators. To my mind, an ongoing challenge in any justice system is the need to ensure that in separating the vulgar from the sacred the design of the spaces of adjudication does not serve to denigrate the public as insignificant.

The principle of open court

Trials apportion blame which justifies the awarding of damages or infliction of suffering and public censure. The implications of exercising these powers require that it is desirable that moral legitimacy should underpin the legal system in the form of public acceptance of process and outcome. This is not an insight we have gained alone; it is a fundamental characteristic of effective dispute resolution across societies and cultures. Successful public figures and institutions have long strengthened their power base when their actions acquire legitimacy through popular support. Adjudicators make decisions which formalise the end of a public dispute but spectators have long rendered the decision credible by acknowledging the standing of the decision makers and inscribing the decision in the collective memory. One reading of Homer's description of the ideal form of adjudication depicted on Achilles' shield discussed in Chapter two, is that the elders were merely presenting a range of considered arguments to the spectators who then decided which was the most convincing amongst them. Legal architecture has long been used to emphasise the importance of proceedings being visible. Sennett (1974) reminds us that in the Greek agora there was a rectangular law court surrounded by a low wall so that citizens going about their business or making an offering to the Gods could also follow the progress of Justice.

Public attendance at courts has also been seen to symbolise the idea that the community had a direct interest in the punishment of offenders. So for instance, Herman (2006) has argued that in the aftermath of the Norman conquest holding courts in the open air was seen as increasingly consistent with the idea that crimes were not just an offence against the victim's family or clan but against the whole community. She suggests that it is only with the signing of the *Magna Carta* that public opinion began to coalesce behind the idea that attendance at a trial was an individual right. The expectation that all those connected with the manor should attend manorial courts and the ability of the monarch's itinerant judges to fine those local Justices of the Peace who did not attend the Assizes are relatively recent practices which also reflect the importance of decision being imbued with community support and the recognition of certain community members as stakeholders in the decision. In this sense it could be argued that the most effective legal rituals are not those which are performed for spectators but those which involve them in a collective experience as participants.

Despite these historical antecedents, for most of the history of the English trial the importance of public justice has been assumed but rarely articulated or explained. Many of the formal bars to secret justice have been born of particular instances of oppression with the court of the Star Chamber being one of the most notorious in an English setting.[1] There are numerous other examples of attempts to close courts to public scrutiny in history.[2] Despite the oft-repeated assertions of experts that the public trial has a long heritage in the common law, Duff *et al* (2007) trace the formulation of a sustained defence of the modern principle of publicity no further back than the nineteenth century to the work of Bentham. His *Rationale of Judicial Evidence*, published in 1827, outlines the many benefits of publicity and the rare exceptions to the principle which should be allowed. Since his foray into the subject, external scrutiny and evaluation of the trial has come to be accepted as an essential part of good legal process and the association between fair trials and public trials. It is now treated as something of a truism that among the many factors contributing to the moral integrity of criminal proceedings is the fact that trials are expected to be held in public places in which spectators unconnected with the trial are able to observe justice being done (Roberts and Zuckerman 2004). In addition the principles enunciated so lucidly by Bentham are now reflected in constitutional documents across the globe. In the West recognition of the importance of publicity emerged most clearly with the codifications of state power that took part at the end of the eighteenth century in America, Ireland and France (Duff *et al* 2007). More recently it has been recognised in the jurisprudence surrounding article 6(i) of the European Convention on Human Rights.[3] The aspiration of open justice is particularly closely bound to the idea of oral testimony in the adversarial trial and has long been cited as one of the reasons why it is considered superior to the inquisitorial system. The emphasis on performance in the adversarial model reflects a belief that trials are not just mechanisms for conveying information to fact finders but are also forums in which evidential narratives which unfold in court can be made accessible and transparent to the public as well. Recent debate about the opening up of family proceedings to the press suggests that the principle of open justice continues to be central to notions of the fair and legitimate trial (see further Brophy and Roberts 2009; Ministry of Justice 2007; DCA 2006).

For many, open justice is treated as synonymous with the notion of a fair and accurate trial because it provides important checks on the credibility of witness testimony and the partiality of the judge. Others have justified the admission of the public to the trial on the basis that it educates spectators in the ways of the law. In their insightful review of these narratives Duff *et al* (2007) suggest an alternative conceptualisation of the matter by promoting the idea of the public

86 Legal Architecture: Justice, due process and the place of law

trial as having independent value. Rather than viewing the public as passive auditors of the trial they suggest a more active role in which open court allows members of the public to make their own assessment of whether culpability has been established and punishment or remedies are deemed fair. Their vision of the public as participating rather than merely spectating means that the public become more than citizens who are being educated or used to provide a check on the behaviour of others. In what may seem like a subtle distinction from other justifications of the open trial, their thesis transforms the public into actors who can evaluate, critique and object to procedure and decisions. Rather than conferring a blanket legitimacy on legal proceedings this approach to the trial imagines that the public are involved in a constant process of conferring or denying the legitimacy of trials and takes us back to the sort of collective experience of law to which tribal moots and community dispute resolution aspired.

Detailed discussions of the role of the public in the trial are rare but there is evidence that those who have attended the public gallery share this aspiration of an actively engaged public in ways other than those emphasised above. Outbursts of angry or distressed supporters are common in certain types of proceedings and can be seen as one manifestation of an active public, though they may result in the protester being removed or prosecuted for contempt of court. Other examples of collective engagement do not necessarily involve disruptive activity. The fact that all those present in the court stand when the judge enters is a simple routine activity which reflects that members of the public are not totally passive in the modern trial. The wearing of badges by supporters of the victim or defendant is another such instance, although in America the wearing of badges with pictures of the dead victim on has been challenged as violating the right to a fair trial.[4] In a more obvious attempt to make clear their engagement with the issues being discussed Radul (2007) suggests that the segregation of the audience turns it into an independent group of performers. In support of this thesis she describes as an example of activism an instance of a trial for the murder of a gay man in which supporters objected to the fact that the court failed to recognise the act as an instance of hate crime. The local press described the incident:

> At one point [. . .] someone said, 'Everone stand in memory of Aaron', and spectators in the gallery, including the victim's family, stood up. Someone shouted, 'This is a crime' but the person who told them to stand said 'Silence please', and they continued standing for more than five minutes [. . .] Judge Humphries did not react to the demonstration.
>
> (Radul 2007, p.8)

Commenting on the case, Radul (2007) suggests that the 'audience' could not literally be said to have been disturbing the court. Instead they staged their own ritual of protest in a zone they perceived as their own.

Keeping the court open

Though not absolute,[5] the fundamental expectation that trials are public should allow us to assume that the public be allocated sufficient space to perform their role effectively and with dignity.[6] In an argument which reflected many of the themes considered in Chapter two, Bentham (1827) has suggested that the holding of courts in buildings severely restricts the possibility of public justice since when courts are held outdoors the only limits on the effective participation of the public are the strength of voices and hearing capacity. Viewed in this way, the wall becomes the primary means by which publicity is limited and denied. However, Herman (2006) warns us against a romantic interpretation of such practices. He reminds us that the holding of courts outside can just as easily be explained by reference to pragmatic criteria:

> One could, as later commentators do, ascribe many functions that would have been served by holding trials and ordeals in public – ranging from the use of such proceedings for their educational or deterrent value, to the desire to ensure that the proceedings be conducted in a manner considered fair – but there is no reason to think that any or all of these *post hoc* justifications actually led to any decision about whether to allow members of the public to attend such proceedings. It seems more likely that the public simply attended such proceedings at will because there were no walls and no reasons to exclude them (p.8).

Whatever the motivation, once it became common for trials to be held indoors, the fact that two or more makeshift courts sat in a large hall contemporaneously meant that the public remained relatively unrestrained outside of the outer bar which marked the edge of a court. Views of Westminster Hall from 1620, the earliest surviving picture of the interior of Winchester Hall from about 1740, and a sketch of Doctors' Commons in 1808 show casual onlookers strolling nearby to the courts and observing proceedings.[7] For Graham (2004) courts during this period are best understood as lively and sociable places. It is clear, for instance that in the Westminster Hall of the seventeenth century the judiciary had to compete with shopkeepers for the attention of the public. Gerhold (1999) shows that merchandise was being sold in the hall by the 1290s and that there were shops selling books, gloves, caps, beer, sugar and linen by 1340. By 1666 there were 48 shops filling the space between courts. In her discussion of the arrangements of the court of King's Bench in Westminster Hall at the opening of Michaelmas term 1485 Blatcher (1978) suggests:

> [The Chief Justice] had to keep his mind on what was happening within his enclosure, which was marked off, as were the other courts, by great oaken planks . . . He must have learned to disregard the traffic of litigants, counsel, attorneys and jurors to the other enclosures and the cries of stallholders and

88 Legal Architecture: Justice, due process and the place of law

> other traders attracted to this concourse of potential buyers, although later when eight 'shops' were erected where their clerks had anciently sat even the King's Bench jibbed. (p.16)

Brookes and Lobban (1997) draw our attention to the fact that the manor court at Aylesford in Kent traditionally met in a building which was open to the street and that passers by often intervened in the proceedings as they went about their business. The fact that trials were much shorter in the pre-Victorian era also meant that groups of litigants were likely to be in the vicinity of the court waiting for their case to be heard. Reflecting on the Assizes in the nineteenth century, Lewis (1982) argues that the courts of the era were very far from being the quiet spaces of today. Indeed, he makes clear that '[t]he public booed and hissed, applauded their favourites, cheered and catcalled'. In part this was because the populace of the time regarded courts as legitimate places of entertainment to which they flooded when there were sensational cases or particular barristers were appearing. Significantly, women and men of all classes continued to attend the court in equal numbers until well into the nineteenth century, despite the ever-present danger of pickpockets. Lewis (1982) describes how in the nineteenth century Lord Campbell occasionally brought his daughters into court to hear the summing up and 'ladies of fashion rushed in to hear the salacious details of the "crim.con" actions blushing prettily all the while' (p.22).

These illustrations and commentaries suggest that not only was movement more free than in modern-day courts but that spectators' behaviour appears to have been much less regulated. The number of people required to attend courts of the period may have made crowd control much more difficult but it is clear that attitudes towards the trial were also considerably different. Reflecting on conduct in the courtroom up until the seventeenth century Prest (1986) has argued:

> The conventions governing the behaviour of judges, counsel and litigants in early modern England plainly differed from those which apply in most Western societies today. Proportionally more people were directly involved in legal proceedings conducted at a much higher emotional and psychological temperature than our modern, purpose built and publicly deserted courtrooms normally witness, except for the occasional cause celebre (p.303).

Up until the end of the nineteenth century illustrations and the amount of accommodation provided by way of public galleries suggests that trials attracted a considerable amount of attention from those not involved in proceedings. Representations of the 'interior' of the courts across jurisdictions commonly show animated spectators in the body of the court engaged in conversation as the trial progressed in the background.[8] Moreover, many of the complaints about courts of the era related to the fact that they were stuffy because of overcrowding (Lewis 1982).[9]

From the late eighteenth century the movement towards segmentation of courtrooms described in chapter three served to facilitate the entrapment of the public

Open justice, the dirty public and the press   89

and others and their separation into distinct groups. As dedicated space was assigned to courtrooms in shared buildings or purpose-built facilities, balconies for spectators became particularly fashionable. In some courthouses such as Carr's Assize courts at York (1773–7) the construction of galleries for spectators also became the means by which areas in a great hall became sectioned off for purely legal use and movement between courts more restricted. Raised galleries for the public are recorded at the courts in Westminster Hall from as early as 1620 and in the Old Bailey Sessions House from 1675. In the case of the latter they also gave court designers additional scope to separate fashionable society from the masses. This is also evident at Nottingham Shire Hall (1769–72) where a balcony with an ornate balustrade was built to accommodate gentile society while other members of the public had to stand at the back of the court. A doorway from the balcony allowed the ladies and gentlemen who sat there during the trial to retire for refreshments with the judge. The fact that certain sections of the public enjoyed more comforts than others is also evidenced by the practice of allowing gentile men and women to sit beside the judge on the bench and enjoy a ringside view of proceedings.[10] In some courthouses access to the public gallery was also restricted by the ability to pay. Bentham (1827) makes clear that courthouse officials in the court of King's Bench charged for entry and at the Old Bailey staff had the right to charge fees for admission to the galleries. It would seem that this was one way in which to limit the high demand for admission. The radical John Wilkes, when Sheriff of London in 1771, thought the practice undemocratic and prohibited it. However, at the October sessions of 1771 there was almost a riot because of the pressure of the crowds trying to get in. Wilkes's order was rescinded as a result and spectators continued to pay to see trials at the Old Bailey until as late as 1860 (Central Criminal Court 2010).

The creation of circulation routes and a backstage and frontstage to purpose-built courthouses described in Chapter three served to further contain and limit the activities of the spectator. Up until the late eighteenth century sociability within the courtroom was encouraged but the move towards specialised accommodation was accompanied by a new emphasis on how the majesty and authority of law could be enhanced through an increasingly managed environment. Significantly this appears to have been accompanied by a tendency to treat the public as irritants who lacked legitimate purpose. In her work on the Palais de Justice in Paris Fischer-Taylor (1993) has catalogued particular concerns about the working classes who were often characterised as being in league with the defendant, or assumed to be thieves who came to court to keep abreast of the law. Fear of an unruly public gallery are evidenced in a number of historic courthouses in England. The wrought iron spikes separating the spectators from the semi-circular inner sanctum of the court at Maidstone, the gargoyles at the doors of Manchester Minshull and sculptures of medieval forms of torture in the public areas of Manchester Assizes did not glorify public involvement in the trial. The latter provided a direct contrast with the statues of the Virtues which adorned the corridor joining the courthouse to the judges' lodgings. 'Speaking walls' in public concourses which commonly repeated bible tracts about bearing false witness, such as those in Victoria Law Courts, also served to remind the public of the gravity of courtroom appearances.

The convenience of judges, lawyers, court administrators and witnesses was undoubtedly served by the creation of discrete circulation routes which led to private offices, retiring rooms and libraries but the public was less well treated in the environs of the increasingly grand temples to justice being built from the late eighteenth century onwards. Significantly, circulation routes provided for the public tended to direct them away from the grand entrance halls on which expenses were often lavished and the public were increasingly required to access the public gallery from a side street through a much less grand entrance. Appleby courthouse (1776–8) has been described as unsophisticated for its time (SAVE 2004) but its central door for the public still led directly to the back of the court, an arrangement which was soon to become established practice for new build. In the courts at Maidstone (1825–6) and Lincoln (1822–8), Robert Smirke reinforced the principle of providing private space for the legal elite by allocating the area which would have normally been used as a central public entrance hall to lawyers and judges and requiring the public to use an arcade on the periphery to access the courts from the back. The Minutes of the Proceedings of the Commissioners for erecting the County Hall in Lincoln show that, having discussed the inconvenience involved in confining spectators 'without' business in large spaces, the Commissioners concluded in March 1826 that only persons of 'rank and distinction' should have access to the central hall of the courthouse. In a similar vein, at Manchester Assize courts whilst the concentric ring of corridors pioneered by Waterhouse ostensibly created a grand communal space, there was actually no need for members of the public appearing as witnesses or the press to wander beyond the 'frontstage' of the building (Mulcahy 2008b). Nowhere is such distaste for the public more forcibly represented than at Liverpool Sessions House (1882) where the elaborate entrances provided for lawyers and judges can be contrasted by the much smaller entrance for the public placed at the back of the building and shown in Figure 5.1. Even when the public gallery was reached, possibilities to observe the proceedings were sometimes hampered. So, for instance, in courtroom number one in the Old Bailey (1902–7) and court one of the former Middlesex Guildhall (1911–14) it is impossible for those in the gallery to see the barristers presenting the case. The result is that the disembodied voices of counsel float up to the gallery and a visual assessment of their credibility as advocates can not be made.

By the time that the Report of the Commissioners for the New Courts of Justice was published in the latter part of the nineteenth century (Royal Commission 1871) the idea of the spectator as someone who interrupts the business of the court had become a dominant narrative. It was asserted by the Commission that most members of the public were not interested in the operation of the law and that those who came to court often represented a frivolous minority. In contrast to visions of the pre-industrial court as sociable,

Open justice, the dirty public and the press 91

Figure 5.1 Entrance for the public at Liverpool Sessions House (1882–4) © Linda Mulcahy.

92 Legal Architecture: Justice, due process and the place of law

considerable emphasis was placed by the Commissioners and those who gave evidence on the need for quiet, order and ample uninterrupted communication. The focus was on providing accommodation for those who had 'legitimate' (p.xxv) business in the new building. There is a clear consensus that separate access to the court should be provided for the 'mere' (p.xxv) spectator and that their passivity should be encouraged. This is made clear for instance in sections of the report which suggest that the public will be less active passive if allowed to sit rather than stand. The Commissioners warned against the provision of circulation routes which would allow members of the public to 'transform the corridors of the building into foot thoroughfares for the district; and [it was suggested] that special footways might be provided so as to tempt the public out of the principal part of the Building' (p.xxvi). Moreover, in their submission to the Commissioners the Joint Committee of the Bar and Solicitors complained that one of the chief inconveniences in court was the presence of members of the public who daily took seats in the court 'not with any view to the particular business before the court, but merely as a place where they can spend their time' (p.64).

In his authoritative work on the Royal Courts of Justice Brownlee (1984) confirms that the new courts of 1882 were largely planned to limit public access. The spectacular cavernous entrance hall was designed by George Edmund Street for use by lawyers rather than the general public who the architect referred to as 'dirty'. This was far from being controversial at the time. The rapid growth of Victorian towns and urban slums together with political unrests reflected in the rise of Chartism and trade unionism meant that the Establishment had much to fear from the new urban working class and immigrants who flocked into the burgeoning industrial towns for work. What emerges in the public debate of the time more generally is an increasing association of the working class with space that was dangerous, ugly, dingy, and disease ridden. As Briggs (1950) has explained:

> The rapid increase in the size of towns in the early nineteenth century amazed and alarmed politicians and administrators. They were afraid of large masses of people over whom they could at best administer a distant control – masses of people who did not fit in the traditional social pattern of small town and predominant countryside (p.67).

The achievements of the industrialists, celebrated amongst other things by the building of symbolic law courts, were matched by a sense of an uncontrolled and unknowable populace; a hungry and rebellious proletariat inhabiting unregulated slums. Viewed in this context the monuments to law constructed in the late eighteenth and nineteenth century are as accurately described as symbols of control and repression as they are monuments to law. One way in which such concerns manifested themselves in the courthouse was in the quest for decorum in the courts. This thesis is borne out by commentary in *County Court Chronicle* in

Open justice, the dirty public and the press   93

which regular column space is given to discussion of the importance of dignity and calm. The early issues of this journal, which was set up at the same time as the new system of County courts was launched, provide an interesting insight into what lawyers of the time considered to be an appropriate atmosphere for judicial proceedings. Editorials and articles reveal something of an obsession with determining what level of ceremony, ritual and surveillance is necessary if the new County courts were to be deemed as credible as the higher courts. A typical correspondent in 1849 gave details of what he considered to be a model County court at Tunbridge Wells:

> In the court the Judge seems rightly to consider his time as belonging to the public, and therefore to every case due attention and time are given. He does not hurry the proceedings (as in the manner of some), as though he were perpetually wanting to dine; and only a reasonable number of causes are named for a fair day's work. The Judge, Clerk and High Bailiff are robed – a proper person is in attendance to keep order, any breach of which is punished as may be requisite. There is a table for professional men who, for the most part, wear their gowns, as they are entitled to do; which they resume, not assume, with the approval of the Judge. Agents (as being wholly unfit to conduct a case) are not allowed to be heard; and, in short those observances are fulfilled, as far as is practicable, which have gained for the superior courts so much admiration for the order of their proceedings in this and in other countries. What is the consequence? Just this – that the suitors entertain and observe the respect due to the administration of the law, and resort thither with satisfaction when it becomes necessary.
>
> (*County Court Chronicle* 1847–9, pp.21–2)

The same correspondent also outlines the obstacles to order and dignity presented by members of the public at one particular trial:

> The suitors and loungers, in various stages of inebriety, talked, laughed, and swore with no interruption other than an occasional 'Pray silence' from the meek-spirited usher; parties came up to the table . . . – the judgment seat! And in more than one instance openly refused compliance with the order of the Judge – such an occurrence seeming so common that no notice was taken of it . . . what is the consequence? Why this, – that the court is derided and abused, its authority disregarded, and when exercised it is called oppression. (p.22)

Concern about decorum and containment within the courtroom had a disproportionate impact on women who were commonly singled out for special comment in this context.[11] Denied the position of lawyer, judge[12] or juror it was the case that for many centuries women could only participate in the trial as litigants, witness or spectators.[13] Although they played a much more active role as litigants than is

94 Legal Architecture: Justice, due process and the place of law

often assumed[14] they were repeatedly denied the same rights as men to embark on litigation (see for instance, Cioni 1985; Stretton 1998; Picard 2003). Despite these bars on full participation, there is evidence to suggest that up until the nineteenth century women were enthusiastic spectators of the trial. But, with the rise of 'gentlemanly values' and the increasing popularity of the view that it was vulgar and debase to show an interest in human suffering, commentators became concerned that an active interest in trials amongst women was unseemly. Undesirable attributes of spectatorship such as morbid curiosity, the courting of attention by flamboyant dress and a tendency to heighten the theatrical elements of the trial by excessive responses to evidence were frequently associated with women. Self control was increasingly associated with masculine intellectual capacity and thus deemed uncommon among women and the restive working class. The inability of women to contain their impulses was a common concern associated with the trend towards female crime and fascination with it (Fischer-Taylor 1993).[15] The English trial with its emphasis on orality and adversarial proceedings was seen as particularly difficult to manage. Those concerned about demeanour in the courtroom become particularly taxed by what Fischer-Taylor (1993) has called the irresistible impulse of the feminine towards disorder and irrationality. Reflecting such concerns, an outraged editor of the *County Court Chronicle* (1847–9) had the following to say of the atmosphere of a scene in Brentford County court:

> The disgraceful confusion . . . is perfectly indescribable; not to be witnessed. It was the whole work, and hard work too, of the poor wicket keeper or sub-bailiff, to prevent regular pitched battles, to say nothing of 'words of violence' to check which his continued entreaty was – 'Silence, ladies! Silence! You really must be quiet, ladies, and go out if you want to talk'. But little the ladies reacted, or gentlemen either, for they still cut their jokes, and vented their wrath, as the humour was upon them, in the most boisterous manner, and this, too, in the immediate presence of the learned Judge, before whom, in several circumstances, the parties after judgement, flatly refused to pay, and threatened their opponents if they enforced it. (p.77)

Other commentators such as Bentham (1827) justified differential treatment of women, and ladies in particular, on the basis that they needed protection from the harsh realities of the public sphere. Characterisations of women as being in need of shielding from the human stories which unfolded in the court were rife in the nineteenth century and the creation of secular divorce courts in 1858 provided a fertile context for more extensive discussion of perceived female sensitivities. Judges regularly ordered women and children out of the court in distressing trials or heard cases involving 'unnatural practices' *in camera*. At the suggestion of Sir Cresswell Cresswell in 1859 the Lord Chancellor proposed alterations to the courts' procedures including the right for the judge to exclude the public whenever the details of the case were of an objectionable nature. When the Bill reached the Commons the Attorney General defended the clause by claiming that women of sensitive feelings, and under the distressing necessity of seeking redress of grievous wrongs, would shrink from having recourse to a tribunal where they would have to relate the story of their husband's cruelty in the presence of a jeering, laughing, and prurient mob. His opponents struck the Bill out but even they applauded the fact that the judge could already send women and juveniles out of court when cases unfit for them to hear commenced. A similar attitude revealed itself in the criminal courts and it was only after they became jurors in the 1920s that judges stopped clearing female spectators from courtrooms when evidence of even a mildly sexual nature was submitted. It will come as no surprise that as a result of such societal pressures women eventually stopped attending the trial in significant numbers in the late nineteenth century (Graham 2004).

Modern-day practices

It has become common in modern times for court architecture to represent law as increasingly democratic; to make attempts to flatten hierarchical structures and move away from the alienating atmosphere of subordination in the courtroom (Melhuish 1996). Policy makers have also identified knowledge of how the courts work as a key aspect of citizenship and some effort is undoubtedly being made to encourage the public to visit the courts in order to better understand their workings. This is especially true in relation to Magistrates' courts[16] (Department for Constitutional Affairs 2003, 2004; Cochlin 2002). Despite these initiatives, modern-day practices continue to suggest an underlying distrust of the public in the courthouse and courtroom. In his study of Wood Green Crown court Paul Rock's (1993) interviews with court staff revealed that the perceived need for segmentation in the environs of the court continues to be fuelled by an overwhelming fear of an unruly and disruptive public. He describes how the fear of being contaminated by the public is reflected in the ways in which staff talk about territory:

> Professionals . . . judge the quality of their home territory by its accessibility to the public. Characteristically, the more open it is, the less it will be favoured . . . Seclusion become desirable when enclosures are surrounded by territory housing the volatile civilian (p.275).

The ways in which fear of the laity dominates the allocation of space is a dominant feature of Rock's work in which he observed ongoing apprehension about the possibility of disorder in the courthouse and its environs. The picture he presents is one in which staff perceived there to be an ever-present fear of the collapse of the social order of the court. By way of contrast he characterises the public in the court as participants with no recognised stake in them and with no knowledge of its languages, topography or inhabitants. He describes the public as entering the court as strangers who waited dejectedly on the benches or in the canteen in a 'morose seclusion' (p.195) and represented the emotional and disorderly.

Despite its democratic aspirations, distrust of the public is clearly reflected in the modern *Court Standards and Design Guide* (Her Majesty's Courts Service 2010). Whilst on the one hand the *Guide* recognises that attending court can be intimidating for the public, more frequent references are made to concerns about graffiti and the potential for vandalism and intimidation by the public. We are informed, for instance, that high-quality materials, railings and plants should be used to deter vandalism, keep people at a distance and discourage loitering. It explains that the justification for the presence of a secure dock in some trials is not just that it protects the public from someone who may be capable of violence but also because it protects the defendant from external attack by the public. The jury, once self informing and located in the body of the court is also characterised as being in need of protection from the public from which they have been drawn. The way in which the principles enshrined in the *Guide* have been realised by architects is also a cause for concern in relation to public facilities. The entrance halls of twenty-first century courts are now shared with the public but the form of the spectator gallery in some courts marginalises the spectator more than ever before. In some Crown court courtrooms such as those at Coventry and Birmingham the spectator sits in a self-contained glass room at the back of the court. They watch proceedings through the reinforced glass of their room as well as through the defendants' glass dock which sits immediately in front of them. Such practices have led Radul (2007) to conclude that the public in these courts are treated like a contained object on display.

Modern-day surveillance of the public is also facilitated through the detailed prescription of who should be able to see whom within the courtroom. The use of height has long been used to facilitate effective participation in the trial and control of it. What has changed since the rise of the centralised design guide is the amount of consideration now being given to 'sightlines' within the court. Indeed, one reading of the *Guide* is that the focus is as much on the visibility of spectators as it is on the visibility of proceedings. The *Guide* provides a particularly good example of Foucauldian accounts of the ways in which law as physical compulsion has been replaced by the simple economic geometry of seamless surveillance. Spectators are expected to have a clear view of the judge but destined to get no more than a 'general view' of the proceedings. Indeed, it is specifically noted that they should have their field of vision restricted. While axial visibility is imposed on them, they suffer from lateral invisibility in ways which clearly associate the public with danger. So, for instance, it is prescribed that those sitting in the public gallery should have the minimum possible direct eye contact with the jury in order to reduce the risk of intimidation of jurors. Moreover, a glass screen between the modern dock and public seating area is expected to be obscured to a height of 1525mm above the floor level so that the public are prevented from seeing the defendant while they are seated.

The positioning of plasma screens to display electronic evidence also appears to have been undertaken without due respect for those in the public gallery in a number of locations. My experience in the course of observing trials for this project has been that these screens are sometimes located so that members of the public get only a partial view of them. Certainly the recent updating of equipment in one courtroom at Snaresbrook Crown court led to a screen which had previously faced the public gallery being removed but not replaced. More recently still, in the video accompanying the launch of the government-funded Virtual court pilot in Kent the camera in court is used to indicate the presence of a public gallery to the defendant but this is not one of the stable images fed through to the remote facility from which the defendants 'participates' in their trial. The result is that the public are rendered invisible to the defendant. These examples raise a number of critical issues which go to the heart of our understanding of the notion of the public or open trial. Since in most courts the only person a member of the public is sure to have a clear view of is the judge it would seem to be the case that the observation of justice is now focussed on observation of the adjudicator rather than evaluation of evidence and the weight which should be afforded it. When evaluation of evidence on screens is limited and close appraisal of exhibits restricted, the idea of the public playing an active role in assessing the credibility of the outcome reached is clearly unobtainable. It would seem that in many of our courts observation has become distinct from participation and viewing from accountability.

Caught reporting

The role of the public in the trial and attitudes towards them are inextricably bound up with the history of the press. As the public have become increasingly contained and silenced in the trial, the press have risen to claim a place as the accepted protectors of the public interest and guardians of the public trial. Histories of the press commonly chart the ways in which reporters have sought to protect the public interest by procuring and disseminating information which the State and powerful private organisations are reluctant to release. Such accounts are apt to focus on the courage of journalists pursuing the right to free speech but punished by fines and imprisonment in their attempts (see for instance, Murray 1972). According to these narratives the interests of the State and the press are in conflict and irreconcilable. A review of legislation over time makes clear that the State has indeed made deliberate and longstanding attempts to stem the growth of the press and in the process silence criticism. The Printing Act of 1695 which required newspapers to be licensed remained in force until the eighteenth century. A stamp tax on every paper was introduced in 1712 and was increased eight times before its eventual removal in 1855 (Barker 2000).[17] Registration requirements were also imposed[18] and libel law was sweeping in its reach, imposed a broad definition of sedition and rendered newspapers vulnerable to legal claims as a result.[19] The impact of such restrictions became all too clear once they were lifted. From only

98 Legal Architecture: Justice, due process and the place of law

a handful of titles at the beginning of the early eighteenth century the newspaper press expanded rapidly until they had become part of everyday life by the early nineteenth century.[20]

In the early years of the press, denial of their role as defenders of the public interest was facilitated by the poor reputation of reporters and up until the eighteenth century scandals about the bribery of journalists appear to have been common (Boyce 1978; Williams 1978). Historians have argued that editors and owners realised that to thrive they must cease to be regarded as a pariah and the claim that they sought to promote as a result was that they would act as an indispensable link between public opinion and public institutions. It was contended that before the public could appraise the value and legitimacy of newly emerged public institutions they had to have information; information which could most easily be supplied by a free press. As a result of its change in focus many members of the press came to be seen as critical in forming and articulating public opinion, defending the rights of citizens and exposing corruption. Indeed, it became commonplace to argue that a free press was vital to the people's right to examine public bodies.[21] Barker (2000) has argued that by the time Walpole became prime minister in 1721 the London press had been accepted as an important and legitimate component of the political system which encouraged the wider population to take an interest in politics and to play a part in it. One implication of their dissemination of information was that the press helped to bring politics out of the restricted arena of the political and social elite and into a much wider public sphere. Although the power of public opinion was still relatively weak in the seventeenth and eighteenth centuries Barker (2000) argues that:

> . . . through the promotion of certain concepts of English liberty, in particular the belief that English men and women were free citizens living in a free state, newspapers encouraged the public to believe they had not just the opportunity, but the right, to involve themselves in the nation's political life and to protest when they disapproved of government action. (p.125)

In the context of the administration of justice this self-assumed constitutional role was translated into the claim that the press should be free to report on what they saw and heard in trials. For most of the seventeenth century, reports of trials were expected to appear only once the trial had concluded and by special permission of the authorities. This position changed as the century came to an end when reporting of Parliament and the law courts was liberalised and from the 1780s onwards press privileges was also extended to correct reports of trial (Grossman 2002). The courts were not to recognise a formal common law right for the press to attend trials until the late eighteenth century and there is considerable evidence of 'skirmishes' about the extent of the right in the years that followed. For instance, press reports of the nineteenth century give instances of a reporter being excluded from the court for making disparaging remarks about the judge and accounts of journalists not reporting proceedings because of their fear of the Judge.[22]

Open justice, the dirty public and the press   99

Whilst it is difficult to discuss the press as a cohesive body it is clear that they have not been shy of drawing attention to their importance in protecting the public interest. In an article criticising the tendency of the courts to try and stifle reporting of divorce cases *Reynold's Newspaper* contended in 1838 that:

> Thanks again to the vigilance of the press . . . attention was drawn to this pro-tected curtailment of the inalienable rights of the public to a fair, full, free, and absolutely open trial of all matter affecting the rights and liberties of anyone. [The courts are] anxious, too, that their frequently pig headed, muddy pated decisions should not be examined at the final bar of public opinion, even police-court magistrates, the smallest legal fry in the world, are beginning to wag their tongues against the press . . . The stronger and more robust journal-ism of the day is not ashamed to let the light of its lantern fall even upon dark places, and make public property that which was performed in secret.[23]

There is undoubtedly some truth behind this rhetoric. The press have played a major role in the revelation of miscarriages of justice. Levy (1967) for instance cites the example of a member of the press alerting the public to the practice of passing judgment in contempt of court cases *in camera* which led to questions in Parliament and a change in court rules. For some commentators, proceedings are afforded greater legitimacy by the mere presence of the press. In the words of the editor of *Reynold's Newspaper* in 1838: '. . . in all matters a magistrate, being human, and therefore likely to err, is likely to more carefully consider his judgements if he knows that they will be weighed up before a great silent tribunal'.[24]

In addition to exposing unfair practice the press has claimed to educate the public about law. The sense of duty to disseminate information in the public inter-est is such that Greenhouse (1996) sees it as an *obligation* of the press as partners of the court in democratic enterprise, rather than a mere right. She tells the story of one modern South African judge, Judge John Didcott, criticising the 'lamentable' performance of reporters who had not thought it necessary to tell the public what the court's reasons for a decision were. The legal press have clearly played a role in unravelling the implications of a number of complex legal decisions not easily understood by the laity. This is particularly important given judges' general unwillingness to discuss their decisions or explain them in lay terms. In some jurisdictions the court provide press officers to explain the implications of a ruling but in England it remains the case that judges communicate to the public solely through their written judgment with the result that it is the press that are left to convey common sense meanings and everyday practicalities. It would seem from such accounts that as the law has become increasingly specialised and the press more confident they have transformed from auditors of the trial to the medium through which the understanding of complex trials is even possible.

With mass circulation of newspapers at the end of the nineteenth century readership of newspapers, and in particular demand for reports of trials, increased

100   Legal Architecture: Justice, due process and the place of law

considerably.[25] The symbiotic relationship between the rise of the adversarial trial and styles of reporting is particularly deserving of mention in this context. Schramm (2000) has argued that in adversarial criminal trials it was eloquent lawyers who were given free rein to weave seductive and eloquent narratives around the evidence procured. In a similar vein, Grossman (2002) identifies the period from the eighteenth to nineteenth century as a period of significant change in styles of reporting and choice of subject matter. Until the mid eighteenth century broadsheet or gallows literature which used formulaic reports and stock characterisation had dominated the market[26] but the supplanting of these by newspapers reflected a new interest in narrative. In particular, he argues that the extended adversarial trial meant that there was much more time to evaluate the defendant as a person and to engage in an analysis of motivation. Accounts of the modern trial were also much better suited to newspapers than broadsheets. Whilst the latter tended to focus on single events and were much better suited to the execution, the modern trial might stretch on for days and weeks and was better geared to serialised accounts in daily newspapers. The arrival and reporting of key witnesses kept interest and momentum buoyant and accounts naturally culminated in the jury's verdict or judges' sentencing. In Grossman's (2002) words: 'When the criminal trial became a battle of lawyers, it also became a newly complex story telling forum' (p.20).

The rise in circulation from the nineteenth century onwards has led Nead (2002) to argue that mass audiences for law were actually created by the press through their transformation of promising legal cases into sensational news. She makes clear that from the 1880s newspapers employed a full battery of innovative journalistic devices including photographic spreads[27] and banner headlines to draw their readers into compelling narratives of desire, betrayal and violence. Trials involving sex or infidelity have proven to be particularly popular subjects (Cranfield 1962).[28] So for example, the *Birmingham Daily Post* reported that in covering Ardlamont divorce case of 1893, the *Scotsman* carried 173 columns over the ten days during which the trial was heard, of which 147 were taken up with verbatim accounts of evidence.[29] Interest in crime and infidelity continues. More recently Jackson (1972) found in his analysis of 100 local weekly newspapers that news about crimes made up nine per cent of main headlines. Of significance for the purposes of this chapter is the observation that as newspaper circulation grew attendance in the public gallery was no longer essential in order to find out about trials. Newspapers could supply many more people with information about trials than any public gallery in the country was able to accommodate. In effect the press were successful in bringing a new kind of 'remote' spectator to the trial. In Nead's (2002) words:

> ... newspapers brought the public into the courtroom; they penetrated its enclosed and rarefied space and opened it up to mass audiences which had been created by the modern press. Every reader had to feel implicated and involved in the progress of the case; they had to feel, through the agency of

the paper, part of the live action that was taking place in and around the court-room' (p.122).

This allowed the press to breech the walls of the court and to bring the drama of the court into the comfort of people's homes. In this sense it was the press rather than the electronic developments discussed later in Chapter seven which led to the first 'remote' trial in which all the participants in a case did not have to be sitting in the same room.

Putting the press in their place

Courtrooms were one of the prime sites in which the claims of the press to take on the mantle of educators and protectors of the public interest were played out and as their status and power increased the press began to call for dedicated space to be allocated to them in the courtroom. Allowing professional representatives of the public some status in the courtroom was clearly attractive to those who feared an unruly public gallery. The agency of the press in ensuring that justice should be seen to be done was suggestive of a legal system eager to be seen to promote the public good but provided a less threatening method for achieving the goal. Viewed in this way, the press could be seen as not only representing the public but becoming the public. As Colinvaux (1967) has asserted: 'In past times judicial proceedings in every country have been before a concourse of people. That was time wasting. Publicity is essential to justice, and the press now performs the essential function of ensuring it' (p.8). Such attitudes towards the press are reflected in the facilities which began to be provided for them in nineteenth-century courts.

Where they once shared observation space with members of the public, the press were gradually allocated dedicated desks within the body of the court. The Crown court at Dorchester (1796–7) provides an early example of this in the form of a fold-down desk at one side of the court. Moreover, when the Assize courts at Durham were refurbished in 1870 the press were given a prominent place in the well of the court adjacent to the witness box and judicial bench. In some instances space for the press was allocated at the expense of that given to members of the public attending the trial. Graham (2004) recounts how when Wyatt constructed an eastern addition to the courts at Winchester in 1871–4 he gave the press ringside seats whilst deliberately restricting the accommodation for spectators. In some locations the press also enjoyed special treatment in the vicinity of the court. *The Morning Chronicle* (1839) provided a favourable account of the insertion of a reporters box above the jury box in the Newcastle Guildhall in the 1830s on the occasion of a local trial that had attracted considerable interest. The fact that a separate entrance was also provided in order to secure a free passage for reporters at all times was particularly praised.[30] Moreover, Waterhouse's Assize courts in Manchester even provided a mess room for the press to use when not in court.

Such facilities were not always easily won. Whilst a number of symbolic courts during the era were designed with the needs of the press in mind some members of

102   Legal Architecture: Justice, due process and the place of law

the judiciary continued to see the press as a threat. This was particularly the case in the lower and provincial courts. The number of complaints made by members of the press to the judiciary about the absence of dedicated space within the court for reporters in the nineteenth century make it clear that they considered their call for dedicated space legitimate by this period but that their view was not always shared by other members of the court.[31] In an article which says much about the status of other spectators in the eyes of the press *Lloyd's Weekly Newspaper* were especially critical of alterations in Clerkenwell police court which placed the press *behind* the public.[32] The pressures on courts built during one era and expected to satisfy the demands of another are well demonstrated in an incident reported in *Freeman's Journal and Daily Commercial Advertiser* in which a dispute arose in the Court of Queen's Bench about who should occupy the space originally set aside for the grand jury.[33] Elsewhere, similar skirmishes seem to have occurred in relation to the press using barristers' boxes when empty.[34] The *Court Standards and Design Guide* (Her Majesty's Courts Service 2010) now lists a press desk as one of the compulsory elements of all Magistrates', Crown or County courts and suggests that minimum provision should be made for two members of the press in each courtroom. In addition, a press room for writing reports and sending copy is listed as necessary ancillary accommodation to be directly located off the main public area and easily accessible from the courts and main entrance.[35] It is noticeable that the *Guide* makes no recommendations as to where the press desk should be positioned within the courtroom but the norm in the Crown court appears to be for it to be located to the side of the tables set aside for advocates.[36]

Few modern studies of court reporting have been undertaken but those that have suggest that although claims for dedicated space in the courthouse are now considered legitimate the claim of the press for dedicated space have not been as successful as other actors such as the lawyers discussed in the last chapter. Indeed, there have been suggestions that suspicions of the press prevail. So for instance, Greenhouse (1996) has commented in a US context that in her experience press rooms are rarely situated in the hub of the court and that access to the judiciary is not generally facilitated. Moreover, in his study of Wood Green Crown court Paul Rock (1993) has observed that despite the presence of a press office set firmly within the courthouse it did not 'convert a stream of agency reporters into insiders' (p.188). Rather, he observed that court staff remained distanced from reporters because they sought to make public what 'insiders' in the court were expected zealously to protect. These findings reflect long-standing tensions between the press and courts which demonstrate that the interests of the former are not necessarily seen as closely aligned to the interests of the court. Certain sections of the press undoubtedly suffer from their tendency to sensationalise and since the nineteenth century crime reporting has at times been condemned as endangering fair trials (Grossman 2002). Cretney (1998) has suggested that concern about enthusiastic reporting of the detail of divorce cases increased significantly with the growth of the popular press in the final quarter of the nineteenth century and prompted considerable debate about the necessity of quelling press coverage of legal proceedings.[37] Commenting on debate around the publication of a photograph of a defendant in the dock receiving the death sentence Nead (2002) suggests that far from seeing such reporting as 'progress' the press took criminal justice back to the streets, to the ballads and rough generic woodcuts of the eighteenth and nineteenth centuries. Concerns have also been voiced that witnesses and jurors might be discouraged from attending the court for fear that their images would appear in the press and they would be exposed to unseen spectators 'present' in unprecedented numbers.[38] Recent consultation papers leading up to the opening up of the family courts to the press demonstrate an ongoing ambivalence about the self-assuming constitutional role of the press as protectors of the public interest. Sensitive family cases which the public are unable to observe can now be viewed by the press at the discretion of the judge but their low level of attendance at the court and the damning assessment of their role made during the consultation period do not suggest that the press are seen as having a particular interest in the balanced dissemination of information about law.

An ambivalent relationship

Rather paradoxically, whilst not wholly responsible for the trend towards discouraging public participation in the trial, the press have been complicit in creating a professional ambivalence about how the trial is rendered a 'public event'. This has largely been achieved through their contribution to the marginalisation and occasional demonisation of spectators at the trial. So for instance, in 1864 it was the press that complained of the public's delight in the '. . . discovery, pursuit and capture of criminals whose offences have gratified a taste for morbid sensation by bringing them within the province of the hangman' (*Preston Guardian* 1864). Nead (2002) has also drawn attention to the various ways in which the press have characterised the public's interest in trials during the early part of the twentieth century as unhealthy. Having created new demands for information about trials there has also been a tendency for the press to then express concern when large numbers of people attempted to gain admission to the public gallery in sensational trials. It would seem that despite claiming the right to protect the public interest the public appeared less attractive in the flesh. The result was that their apparently insatiable appetite for first-hand experience of trials was repeatedly conceptualised as undignified.

The newspaper report promised to allow the public to subvert the behavioural code of the modern courtroom by consuming trials from the comfort of their home. In this way, it was the photographs of trials and courthouses which appeared in the illustrated newspapers of the late nineteenth century which began to mediate between official rituals in the enclosed spaces of law and the private sphere of the home. But the new interest in attending trials which it encouraged was not so well received by the press and suggests that their role as disseminators of information was closely guarded. In a curious twist the resurgence of public interest in the trial became an additional element in the stories being told

104   Legal Architecture: Justice, due process and the place of law

(Nead 2002). The group singled out for special comment by the newspapers employing a now familiar discourse about them, were the female spectators who were considered to be morbid, cruel and sensation loving. In Nead's (2002) words:

> It was the women, of all ages and different classes, who were defined as the voracious consumers of murder. Women queued overnight, they scrambled for admission and packed into the courtroom. Their appetite for murder seemed to call into question the very nature of modern femininity (p.123).

There are a number of things particularly worthy of note about these reports. Firstly, the many women caught up in the trial as defendant, witnesses, spectator and newspaper reader are treated as indistinguishable. Their inability to distinguish between 'good' and 'bad' women clearly worried and perplexed the press, especially when both female defendants and female spectators appeared equally as respectable. The reports also reveal intense concern about women transgressing the boundaries of the courthouse. Women spectators were criticised as spilling over into the precincts of the court and public highway as they queued for places in the public gallery. Indeed, the inability of the building to contain them was often seen as a public order issue.

The reinvention of public space

The filming and broadcasting of trials has the potential to play out this debate in new ways and to open up debate about aspirations to an open court once more. The broadcasting of trials on the internet and through television networks offers a modern-day solution to the problem of restricted space within courthouses as well as the problem of low turnout in the public galleries of today. In support of the controversial use of cameras in the OJ Simpson trial Judge Arnold, Chief Judge of the US Court of Appeals has argued:

> Maybe I'm just stubborn, but my general feeling is that the public owns the courtroom. They pay for it, it's part of their government. And the more they know about what goes on there, the better, in the long run, the courts will be.
>
> Arnold and Merritt (1999) p.85

Interest in such initiatives is such that in the US the televised coverage of trials has been labelled a growth industry fuelled by demand for publicity from judges,[39] the oversupply of lawyers competing for clients and a public eager for courtroom drama (Lassiter 1996). In an English setting photographers have been banned from the courts since 1925[40] but in Scotland, cameras have been allowed inside courtrooms since 1992 albeit under strict guidelines.[41] In August 2004, Lord Falconer announced a six-week pilot scheme to film a small number of appeal court hearings at the Royal Courts of Justice in London on the basis that the issue was 'ripe for a debate' and almost all the proceedings of the new Supreme Court are now televised though not 'streamed' or easily available. Despite lobbying from all the major broadcasters for an announcement about a general policy to be made, the Ministry of Justice has still yet to make any further progress. However, support for cameras in court appears to be gaining momentum. Keir Starmer used his first major speech as Director of Public Prosecutions to come out in support of allowing television cameras into trials arguing that it would bring 'a breath of fresh air' to the criminal justice system (McNally 2009). Pressure to pursue this goal is also coming from the media. Sky have recently launched a campaign to allow cameras into courts to film proceedings on a routine basis (Gibb 2010).[42]

Practices in relation to the filming of trials vary across jurisdictions but judicial resistance to the practice appears widespread. Opponents of the idea have argued that the transmission of trials disrupt proceedings, undermine the rights of those involved in trials who object to the practice and encourage counsel to play to the cameras. It is around these issues that extensive debate has been conducted in the United States.[43] Excessive publicity from OJ Simpson's trial for murder, possibly the most watched event in history (Lassiter 1996) has prompted some reticence about introducing TV cameras to the court across the globe.[44] Spatial indulgence of the press in the trial was such that Sanford (1999) was moved to describe the press as roaming the courtroom 'like proprietary floorwalkers' (p.4) and a considerable amount of adverse commentary focused on the ways in which such sensational trials can lead to a media 'circus'. Others have expressed fears that reporting will not present a balanced approach to proceedings. In an empirical analysis of how five television news outlets in New York used open access to this trial researchers found that of the 12 hours spent covering the trial; 65 per cent of the coverage did not show video footage from inside the courtroom and 79 per cent of the coverage did not air audiotape from inside the courtroom (Pogorzelski and Brewer 2009).

Even when one takes these concerns into account in the media-savvy twenty-first century it is difficult to understand how objections to the televising of trials can sensibly continue. It is not easy to chart the reduction in numbers of spectators who attend the public gallery with any accuracy, but what is clear is that public attendance in the trial has reduced significantly over time. Historic courthouses make this shift all too clear. Whilst nineteenth-century courts could commonly accommodate almost 200 people standing, modern courts commonly have no more than 25 seats in the public area of the standard court. Writing in 1842 Joseph Gwilt was so incensed by this trend that he suggested that architects of courthouses should concentrate on avoiding the common error of paying too much attention to judges and barristers while leaving everyone else in the courtroom penned up like poor sheep at Smithfield (SAVE 2004). There are undoubtedly a number of factors which have contributed to the demise of observers such as changing working patterns and the widespread availability of television as a form of entertainment. In this chapter I have suggested that the design of courthouses and attitudes to the public expressed by those commissioning and designing them has also played a major and largely uncharted role in discouraging use of the public gallery. However, it remains important to stress that a lack of attendance in court does not always reflect a lack of interest in the justice system, a fact to which the popularity of courtroom drama attests. Moreover, whilst physical attendance at court may not be as attractive as it once was the popularity of television and the internet provides policy makers with the ideal opportunity, should they wish to take it, in promoting interest in the workings of the justice system and addressing concerns about the democratic deficit.

Concerns about the media circus are credible when one examines the miscarriages of justice caused by over-reporting of trials and lack of dignity afforded proceedings. The public have also played a part in the hysteria which surrounds such events. Certain types of trial have always attracted high levels of attention and possibly the justice system would not be working properly if they did not. It is also the case that media participation in cases can be managed more effectively than was the case in the OJ Simpson trial. One approach has been for courts to maintain close control of the transmission and production of film clips and photographs. This is the practice of the international criminal court which films and broadcasts its own proceedings with a 30 minute video streaming delay.[45] Perhaps most importantly for those interested in the spatial dynamics of the courtroom on which this book focuses the value of the televised trial is the way in which it provides an opportunity for viewers to make their own assessment of evidence without the constant surveillance of court staff. The fact that the viewer of televised proceedings controls the space from which they watch the trial and the length of that engagement can be understood as a liberating experience for those of us used to the constraints and occasional degradation in public galleries in supervised courts. A final justification for a more liberal approach to the televising of the trial is that objections to it can only appear hypercritical when one looks at the enthusiasm with which the government and judiciary have introduced live link for defendants in routine criminal trials. Observers can legitimately pose the question of what makes it acceptable to allow the evidence of a defendant to be conveyed into the court through video link but prohibits the transmission of the proceedings to the public in whose name the case is being held.

Conclusion

In this chapter and the last I have argued that the use of space within the courtroom tells us much about the ideologies underpinning judicial process and power dynamics in the trial. In particular I have sought to explore the ways in which progressive use of segregation and surveillance to contain participants have served to undermine any meaningful aspiration to participatory justice or notions of the courtroom as a public space. The picture painted here is in direct contrast to many of the statements issued by the Ministry of Justice about the courthouse as an increasingly democratic space. There is undoubtedly evidence of attempts to make courthouses and courtrooms more accessible to the public but there is also

Open justice, the dirty public and the press   107

considerable evidence of historical spatial practices rooted in pre-democratic societies which deserve to be questioned.

It might also be argued that although the courtroom has become flatter in recent decades with less emphasis on overbearing symbols of power there is also a possibility that discipline and surveillance in the courtroom has become so subtle that these crude symbols of force can now be dispensed with. For all its talk of accessibility producers of the modern *Court Standards and Design Guide* (Her Majesty's Courts Service 2010) can remain confident that they have perfected models of containment unconceived-of by their forebears. The complete segregation of clearly defined categories of participant in the trial, the creation of private zones within the courthouse and courtroom, detailed specifications as regards sightlines and the physical separation of the press and jury from the public are all architectural embodiments of control in which notions of 'visibility' could be seen as a ruse. Contrary to the rhetoric employed by policy makers the architectural apparatus imposed by the *Guide* can just as easily be read as a vehicle for creating and sustaining power relations as it can a site where equality is valued. The sophisticated forms of segregation and surveillance employed allows things to be arranged in such a way that the exercise of power is subtly present in ways which increases its efficiency and transforms spectators into docile spectators.

Notes

1 Controversy about the secrecy of proceedings continue to this day. Two examples which have attracted considerable attention are the Court of Protection and the Special Immigration Appeals Commission.
2 So for instance, the leet court of the honour of Raylie in Essex met at night without light on a hill outside the town. No notice of the proceedings was given as members of the community were expected to remember the day of the year on which the court always sat. Once the participants were assembled the steward called the names of the suitors in as low a voice as he could (Brookes and Lobban 1997).
3 This declares that everyone facing criminal charges is entitled to a public hearing.
4 See for instance *Musladin v Lamarque* (2005) US Court of Appeals 9th circuit, No 03-16653.
5 Bentham (1827) suggested several circumstances in which privacy of proceedings could be justified. These included the interests of order in the courtroom, the reputation and tranquillity of families involved in proceedings, public decency and state secrets. To this list Duff *et al* (2007) have added the prevention of pre-trial publicity which could prejudice a fair trial, the identity of certain witnesses and proceedings relating to juveniles. It was in the courts of summary justice that the principle should be seen to finally break down. The Children Act (1908) required that offenders under 16 be tried in a different place or at least at a different time from the regular court. It excluded spectators from these occasions although the principle of public accountability was maintained by permitting the press to attend. As in 1859 the principle of open court was challenged in the name of the family. The 1908 Act formally banned children from watching any criminal trial.
6 Other factors which Roberts and Zuckerman (2004) list as contributing to the moral integrity of criminal proceedings include an independent judiciary, respect for citizens' rights and proof by reliable evidence.
7 These pictures can be seen in Graham (2003).

108 Legal Architecture: Justice, due process and the place of law

8 See for instance the illustration of courts reproduced by Graham including view of the Court of King's Bench, Westminster Hall from about 1755, the drawing of the Court of Common Pleas in Westminster Hall built in 1740–1, the sketch of the imaginary reconstruction of a trial in the Earl Marshall's court at the College of Arms circa 1707 and the courtroom at Bow Street dated 1808 all reproduced in Graham. The Court of Chancery circa 1725 reproduced in Gerhold (1999) also illustrates this point as does an engraving of the Court of Requests of the City of London by Robert Wilkinson from 1817 reproduced in Herber (1999).

9 Lewis (1982) makes reference to the fact that ventilation became such a problem at Lincoln Assizes in 1852 that the judge ordered that the windows be opened. When it was discovered that this could only be done if someone climbed on to the roof he ordered that the glass be smashed instead.

10 I was informed during a tour of the court by a guide that in a wonderful twist of fate the balcony also housed the very vocal wives of miners prosecuted during the Nottinghamshire miners strike.

11 The issue of whether space is gendered and if so *how* it is gendered is a problematic one which involves an analysis at a number of different levels. Jane Rendell (1999) has suggested that a whole series of questions can be asked about whether gendered space is produced through intentional acts of architectural design, according to the sex of the architect, through the interpretative lens of architectural criticism and through use.

12 One exception is worthy of note. Women who served on the jury of matrons could act as adjudicators in the trial. The Jury of Matrons was a form of special jury which was used to resolve legal disputes over whether or not a party to a legal action was pregnant. Pregnancy could prevent the property of a recently deceased man passing to a beneficiary. A hanging might also be delayed if the prisoner was pregnant. The jury of matrons survived until the early nineteenth century (see further Oldham 2005).

13 The first female solicitor, Carrie Morrison, was not admitted until 1922; the first woman was not called to the bar until 1921 and the first female King's Counsel was not created until 1949. It is worthy of note that males still make up 71 per cent of self-employed barristers. This is despite a nearly even split of male and female entrants to the bar (*The Lawyer* 2009). It has taken much longer for women to be allowed to perform adjudicatory functions. The first female juror was sworn in at the Old Bailey in January 1921. Part of the rationale behind their exclusion from the jury was explained by Sir William Blackstone in the nineteenth century. He opined that the common law requires jurors to be free and trustworthy 'human beings', and that while the term 'human beings' means men and women, the female is, however, excluded on account of the defect of sex (Matthews 1927). The first female County court judge, Elizabeth Lane, was appointed in 1962. But it was not until six years ago that a woman occupied this space in all courts when Brenda Hale became the first female judge in the House of Lords. There was considerable excitement in the press in October 2008 when record numbers of female judges were appointed. Five of the 22 judges selected for the High Court in that year were women, taking their numbers to 17 out of 110. Even with the new additions, men will make up nearly 85 per cent of senior judges. Patricia Scotland was the first black woman to become a QC in 1991. See further online. Available HTTP: <http://www.bbc.co.uk/radio4/womanshour/timeline/2000.shtml>

14 Picard (2003), for instance, has asserted that in the 1570s plaintiffs in half the cases brought before the ecclesiastical or bawdy courts which dealt with such matters as marriage disputes and sexual assaults were women, as were half the witnesses.

15 See also Douzinas' (1992) thesis about law's general fear of the image.

16 See online. Available HTTP: <http://www.magistrates-association>

17 For an excellent account of newspaper stamps one should look at Chandler and Dagnall (1981). The first stamp duties were imposed by the Stamp Act of 1694 in the reign of William and Mary, although the most discussed piece of legislation is the Stamp Act

Open justice, the dirty public and the press   109

of 1712. See also Newspaper Stamp Duties Act 1819. Stamp duty was increased by 266 per cent between 1789 and 1815 and publications subject to the stamp duty were redefined in 1819 to include political periodicals (Curran and Seaton 1997).

18 The result was that at various times publishers of weeklies had to lodge details of their papers and place financial bonds of between £200 and £300 with named authorities (Curran and Seaton 1997).

19 For all of the eighteenth and the first half of the nineteenth century truth was no defence to an action although a change to the Libel Act in 1843 allowed a defendant to plead that what they had written or published was in the public interest. In the 1792 Libel Act Fox secured the right of the press to criticise politicians. The Act left it to juries not judges to determine whether an alleged libel could be held seditious. In addition, all those involved in writing, printing, distributing and selling material could be charged when what was published was likely to bring into contempt or hatred the ruler, his heirs or successors; government and great national institutions or cause disaffectation amongst them. Liberation from these regulatory constraints came in a number of forms including the abolition of the Court of Star Chamber 1641 which had become, by the seventeenth century, an instrument for the control of the press which exercised a tyrannical control over the embryonic newspapers of the day (Murray 1972). The ending of press licensing in 1694–5 came about as a result of much criticism of licensing including John Milton's pamphlet *Areopagiticia*. Press taxation, also called the tax on knowledge, was repealed 1853–61. See also the Libel Acts of 1792 and 1843. For a history of censorship in Elizabethan England see Clegg (1997).

20 The population of England also grew rapidly between 1695 and 1855 leading to more demand for printed materials. This was coupled with a consumer revolution spurred on by the nation's increasing prosperity. Finally great advances in technology allowed for greater output. So for instance steam printing of *The Times* began in 1814 (Williams 1978).

21 In parallel with this have run concerns about the potential for the press to delude the public if used corruptly (Barker 2000).

22 *County Court Chronicle* (1854); *The Morning Chronicle* (1840); *Freeman's Journal and Daily Commercial Advertiser* (1848). See also the report in an Ulster newspaper in 1898 that reporters were ordered to leave the court when a Magistrate decided to hear a case of indecent assault *in camera*. The paper commented that the Magistrate had assumed powers not claimed by a judge of the High Court who had recently asserted that he had no power to ask the press to retire, 'Petty Sessions' *The Belfast Newsletter*, 3rd September 1898, Issue 25922.

23 *Reynold's Newspaper* (1838).

24 *Reynold's Newspaper* (1838).

25 As early as 1732 the *Derby Mercury* began to print an account of the most remarkable trials at the Old Bailey. It is difficult to know with any certainty who was reading papers in the eighteenth and nineteenth centuries but historical sources suggest that one paper might be read many times and that readership was not confined to the upper and emerging middle classes. Contemporary accounts suggest that during the early eighteenth century workmen habitually began the day by going to coffee rooms to read the paper or clubbed together to buy one. Analyses of the actual texts of papers including the advertisements placed there suggests that newspapers largely remained the concern of the growing middle class but from the mid eighteenth century onwards papers directed at the lower orders such *as The Universal Chronicle, Reynold's Weekly Newspaper, Political Register*, or *Lloyd's* began to be produced. There is considerable debate about levels of literary over this period but it is clear that dissemination of the news and arguments contained in newspapers was facilitated by the fact that many newspapers were read aloud and that this was something expected of the literate who associated with the illiterate. The practice was so widespread that many landlords of

pubs would pay readers to be present. The eighteenth century also witnessed the rise of libraries and subscription reading rooms (Barker 2000). Foreign commentators of the time suggest that newspaper reading was a particularly English passion.

26 Gretton (1980) notes that the reader's attention was often drawn to this when the picture and report did not tally. As an example Grossman (2002) notes cases of illustrations which had the wrong number of people being hanged. Gretton (1980) suggests that of all the different types of broadsheets produced the most popular, the murder sheet, was generally the most conservative. When depicted trial scenes were illustrated very schematically and were essentially moralistic tracts which rarely protested against the offender's condemnation and tended to present just one response to the punishment of crime. He has suggested that this reflected the ways in which highly ritualised trials and executions convert the unique and unfamiliar into the familiar and recurrent.

27 Before photography within the precincts of the court was banned photographs of major criminal trials reflected a shift towards a new journalism which relied more than ever before on the image as news transmission moved from a predominantly textual to pictorial form (Nead 2002).

28 Cranfield (1962) cites as examples the trial of Miss Blandy in 1752 for the murder of her uncle which took up a whole page of the *Cambridge Journal* for a month and the affair of Elizabeth Canning in 1753 who appeared to have been kidnapped by a brothel owner. But even in the absence of high profile and sensational cases crime remained an essential ingredient of the eighteenth-century newspaper.

29 They estimated that the descriptive matter contained upwards of 52,000 words and the verbatim accounts 346,000. Further it was asserted that the General Post Office telegraphed 186,000 words about the case daily. Upwards of 100 pressmen were engaged in reporting the trial including 70 reporters, 21 descriptive writers and 15 artists. As it was impossible to accommodate all of these people in the court a system of relays was adopted. *Birmingham Daily Post* (1893).

30 *Morning Chronicle* (1839).

31 See for example *The Belfast Newsletter* (1884) and *The Western Mail* (1886).

32 *Lloyd's Weekly Newspaper* (1869).

33 The County Sheriff had come to an arrangement with the press that they should occupy it. Supporting the claims of the press Judge Burton negotiated an accommodation whereby the press would agree to vacate the space if members of the Grand Jury chose to use it. The Grand Jury also had a balcony from which they could address the court. See further *Freeman's Journal and Daily Commercial Advertiser* (1844).

34 See for instance *The Times* (1848).

35 Section 17 of the *Guide* indicates that the room should contain a pay telephone, a voice alarm and should be linked into the public address system.

36 This is confirmed by a diagram in the *Guide* at page 196; the same document suggests that in typical youth courts the press sit at a separate desk towards the very back of the court (p.159).

37 Before the creation of the divorce court in 1857 proceedings in ecclesiastical courts which dealt with many family matters has been held in private. The Matrimonial Causes Act of 1857 required witnesses in the new divorce court to be examined orally in open court and a practice developed whereby crowds congregated to hear the detail of marriage breakdowns. The King's expressed disgust at the reporting of another case led in due course to the Judicial Proceedings (Regulation of Reports) Act 1926. The effect of the Act was to prohibit the publication of indecent matter and to stop the press from giving detailed verbatim accounts of sensational divorce cases.

38 So, for example, The Children and Young Persons Act 1933 provides that no newspaper report of proceedings in a juvenile court shall reveal the name, address, school or other identifying characteristics of a person under 18 against whom proceedings are taken or a witness in the case. The ban will only be dispensed with if the court or

Secretary of State thinks that the interests of justice require it. The Criminal Justice Act of 1925 provided the official response to such practices, the growing power of commercial popular culture and the tyranny of the press. Under section 41 of the Act the making of an image of a courtroom and its precincts and the publication of the image is an offence. It is also the case that the press have been restricted in their right to report the full details of trials involving juveniles of official secrets. See also Official Secrets Act 1989, Security Service Act 1989.

39 Although Lassiter (1996) has also claimed that such initiatives are less likely to be supported by Federal Judges who do not rely on high visibility for election.

40 Section 41 of the 1925 Criminal Justice Act reads: 'No person shall take or attempt to take in any court any photograph, or with a view to publication make or attempt to make in any court any portrait or sketch, of any person'.

41 See online. Available HTTP <http://www.bbc.co.uk/blogs/theeditors/2008/05/cameras_in_court.html> (Accessed February 2010).

42 The Chillcott inquiry was televised and the BBC was given permission to film the proceedings of the appeal by Abdelbaset ali Mohmed al-Megrahi against his conviction in the Lockerbie case but only after long negotiations.

43 In 1965, the U.S. Supreme Court appeared to say that the presence of cameras could be detrimental to the defendant's case by undermining due process under the fourteenth amendment *Estes v Texas*, 381 U.S. 532, 85 S. Ct. 1628, 14 L. Ed. 2d 543 but later changed its mind in the case of. *Chandler v Florida*, in 1980, a case which spurned a raft of permissive state legislation as technology improved and became less intrusive. *Chandler v. Florida*, 449 U.S. 560, 101 S. Ct. 802, 66 L. Ed. 2d 740.

44 See, for instance, the judicial opinions expressed online. Available HTTP: <http://www.c-span.org/camerasinthecourt/> (Accessed February 2010).

45 Online. Available HTTP: <http://www.icc-cpi.int/Menus/ICC>

Chapter 6

The heyday of court design?

Introduction

This chapter and the next turn to look at two particular periods in the history of court architecture and design in an attempt to unravel what the state of architecture can tell us about the weight attached to the administration of justice in society at different times. In this chapter I focus on the revolutionary transformation of courthouse design from buildings located in castles, market halls and shire halls to the purpose-built and monumental designs of the long nineteenth century.[1] The ambitions of those commissioning and designing courthouses of this period is such that this era can be best understood as marking the heyday of architectural ambition for courts in England. It is also during this period that modern templates for design began to emerge. Indeed many of the characteristics of contemporary courts can be traced to this period. In some ways the story I tell in this chapter is unexceptional. Many of the changes in the architectural style which occurred in the late eighteenth and early nineteenth century can be explained by the development of specialised building types more generally. A burgeoning interest in style, a demand for greater standards of comfort, greater wealth and the emergence of new concepts of public and private spheres all led to a demand for new, grander and increasingly differentiated buildings. Despite this, I argue that there is also a more nuanced account of the development of the law court as a specialist building type which deserves to be told and has been largely neglected to date. It is a story about the inextricable link between the modern law court, the process of industrialisation and the emergence of a much-reformed legal system. In the chapter which follows I turn to compare the ambitions of the nineteenth century with thinking about court design today.

The story of the new town halls, concert halls, reading rooms and libraries planned and often funded by industrialists in the nineteenth century has been charted extensively elsewhere. But despite the extensive literature on Victorian architecture very few writers have turned their attention to the role of the law court in expressing the new ideals of the market place and public sphere. When they are discussed, law courts tend to be included in lists of new specialist building types or as examples of Victorian ambition and civic pride. Graham (2004)

The heyday of court design? 113

alone hints that it is during this period that the scale of provincial courts and the resources lavished, or even 'squandered' (p.267) upon them can not be understood by reference to general trends alone. In this chapter I explore this assertion in more depth and look at the association of 'new money' with the development of a particular model of the law court. I argue that the links between a reformed legal system which served the mercantile classes and the building programmes they championed served to aggrandise the role of law in modernity. It is in this context which we can best understand the ways in which the law court was transformed from a shire hall to a building which was particularly deserving of recognition and celebration in the new civic landscape. I argue that three particular social and political movements were to alter the spatial practices of law from the late eighteenth century onwards. At national level, an appetite for change, prompted in part by a new radical politics worked as a major driver of centralised legal reform and the undermining of a largely local legal system based on privilege. In turn, this was to lead to new hopes for law and a reformed legal system as an instrument of social change and bulwark of a rational society in which success was based on merit rather than privilege. Finally, I contend that it was the mobilisation of law fuelled by a new class of merchant princes keen to glorify their own achievements which provided the most pressing imperative in courthouse design over this period.

A taste for comfort

Throughout the history of courthouse design there have been examples of beautiful, extravagant or grand buildings which stand out from the norm. Courts such as those built at Derby (1659–60), Northampton (1675–8) and Abingdon (1677–82) undoubtedly set high standards for Assize accommodation in their time (see Figure 2.2). Better standards still were established at York (1772–7), Hertfordshire (1768–70) and Middlesex Sessions House in the century that followed. But wide variation in standards was the norm. To a large extent this is a product of the way in which the building of new law courts was funded. It was not until the 1970s that the responsibility for design and funding of courthouses was assumed by central government. Buildings which were used to house the courts before then were often commissioned by benefactors of a town, be it a member of the local gentry or Guild and commonly served as a monument to their generosity and power. In other cases courthouses were funded by a charge on the local rates and the funds available reflected the resources of the local community. The variations in style and grandeur produced by this 'system' of funding can be illustrated if we compare the Guildhall at Bath rebuilt from 1775 onwards to that of Appleby courthouse built between 1776 and 1778. While the former contained a banqueting room, council room, court room, mayor's parlour, three rooms for his clerk and assistants, a records room, surveyor's office, kitchens and living accommodation for the caretaker (Girouard 1990), the latter consisted of just two simple rooms for use by the courts and borough and has been described as justice stripped to its basics (SAVE 2004).

114   Legal Architecture: Justice, due process and the place of law

Buildings of a very high architectural quality remained relatively unusual in the eighteenth century.[2] Indeed, it is clear that in some places standards were very low.[3] So, for instance, as late as 1784 a temporary shed for the Assizes was erected against the ruinous castle walls in Buckingham. Moreover, prior to the opening of the new courthouse in Guildford in 1862 the Crown and nisi prius courts had been held at opposite ends of the street and one of the courthouses was used for markets at the same time as trials. *The Law Times* (1862) recounted one instance of the judge having to ask for the market to be cleared because he could not follow what was going on in proceedings. The unsatisfactory standards of some accommodation for courts is also potently illustrated by the standards of the Royal Courts in London. Colvin's (1977) *History of the King's Works* suggests that throughout the seventeenth century the makeshift wooden enclosures which sectioned off the courts of King's Bench, Chancery and Common Pleas from the rest of Westminster Hall were inconveniently situated, draughty and uncomfortable. King's Bench and Chancery were placed behind a new gothic screen in 1739 and roofed over in 1755 while the court of Common Pleas was moved to a new building near the hall in 1741. But even these arrangements proved unworkable. In 1821 the courts in the hall had to be demolished so that the space could be made available for the coronation banquet. Maitland (1883–4) reported that there was also general outcry against the plans which John Soane[4] drew up for replacement courts in the early nineteenth century because there was no room for the bar, the press, the public, libraries, coffee rooms and waiting space save for in the chilly hall. When constructed, the new courts were also immediately considered inadequate and highly inconvenient for those using them and it has been claimed that dignity was regularly sacrificed as a result (Brownlee 1984; Port 1968). In the 1860s the second court of Queen's Bench was closed because the Chief Justice declared that it was detrimental to the health of all attending. The result was that an exodus to other locations began almost immediately and in 1849 the equity courts took up permanent residence in Lincoln's Inn.

By the end of the eighteenth century a steady demand for sole use courthouses, in which lawyers did not have to share buildings with other public functions began to emerge. But even where ambitious and well-funded projects were conceived, the paucity of templates for design determined that architects of the period had very little to guide their planning. The result was that new build was born into something of an architectural vacuum (Watkins 1982). By the beginning of the nineteenth century those few purpose-built court buildings and courtrooms which did exist were not always considered suitable models for the more ambitious projects being now envisaged. In 1811 the architect Daniel Alexander boldly commented that there were very few examples of courts which could with propriety be copied (Chalklin 1984).[5] When new accommodation for the Royal Courts of Justice in the Strand[6] were being conceived the assistance of the Foreign Office was enlisted in procuring plans of foreign courts but the Commission set up to oversee design concluded that few examples found could be of service unless it was what ought to be avoided (Royal Commission 1871; Port 1968). In the event

The heyday of court design? 115

57 paragraphs and 66 schedules were prepared for the architectural competition. This voluminous anthology (Brownlee 1984) reflected the desire for a much higher standard of accommodation than had ever been evidenced in the courts based at Westminster Hall or indeed anywhere else in the nation.

It is clear that, in part, developments in courthouse design over this period mirrored more general shifts in demands for new types of architecture. Whilst churches and palaces remained the main architectural exhibits of the first half of the eighteenth century, the second half saw a number of new types of buildings being constructed for use by the public.[7] Many of these such as schools, mechanics institutes, museums, exhibition spaces, reading rooms, libraries and art galleries represented the physical expression of the ideas of the Enlightenment and the newly evangelical quest for self improvement. Others such as railway stations, warehouses, factories and banks were born of industrialisation and the movement towards urbanisation which followed it. By 1851 there were more people living in towns than in the countryside and the towns grew rapidly as ever more houses, factories, shops, churches, railways and docks were constructed. Large connurbations had also developed during previous centuries but it was the scale and speed of construction which was unknown and hundreds of architects, engineers and surveyors were needed to plan and execute this work (Cunningham 1985).

In addition to the emergence of new building types for which there was no architectural precedent, buildings were also being constructed on scales never imagined in the past (Pevsner 1976). Whatever their heritage or function the desire to impress united the best examples of each building type of the era. In the words of Briggs (1993) the Victorians:

> ... liked 'imposing' buildings with 'pretensions'. They loved symbolism. They were seldom afraid of exuberance. Individuality and status seeking were both expressed in their villas: their banks and shops became more and more ornate and decorated as the century went by. Even mills, warehouses and docks, which retained a dignified and still handsome functionalism throughout the early years of the industrial revolution, became places to decorate as well as to use. Their churches might be in many styles, for they ceased to be dominated by rules as to which style was appropriate for which kinds of buildings (pp.45–6).

It followed from these aspirations that public buildings, including the courts, came to be seen as major architectural commissions. Many of the greatest architects of the late eighteenth and nineteenth century such as John Carr, John Soane, Charles Barry, George Edmund Street, Gilbert Scott and Alfred Waterhouse were involved in courthouse design and a small group of architects began to take responsibility for the design of more than one courthouse. Though largely unacknowledged in accounts of his life and work, by the time of his death Robert Smirke had designed more buildings to house courts than any other architect before him.[8] As standards and expectations rose the courthouse emerged as a specialist branch of design and this trend was reinforced by the emergence of new organs of local government. Surveyors began to oversee the building of the new County courts from the mid nineteenth century onwards and police surveyors started to supervise the building of the new police courts around the same time. Both initiatives lead to increased continuity of design within and across districts.[9]

Competition to host the Assizes was often fierce and prompted ever higher aspirations when it came to design. Pressure for improvements often came from the judiciary and was backed, implicitly or explicitly, by the threat to hold the Assizes elsewhere or to fine the County.[10] Horsham Town Hall (1721) was partially rebuilt in 1809 at the expense of the Duke of Norfolk following judicial threats to move the Assize and Quarter Sessions to better accommodation at Lewes. Longstanding competition over Assize accommodation also took place between Abingdon and Reading in Berkshire, and Aylesbury and Buckingham in Buckinghamshire. In addition to prestige, the revenue generated by the Assizes was also a considerable incentive to improve conditions. The arrival of the monarch's itinerant judges heralded a succession of balls and assemblies in their honour and local trade was also boosted by members of the public who came to spectate. Officials in the town of Salisbury calculated that during this period those attending the Assizes brought £10,000 income in the town each year.

These various factors meant that from the late eighteenth century there was a general trend for courthouses and shire halls to be built to a much higher specification than their predecessors.[11] Commentators continued to complain about the conditions of courts throughout this period and the century which followed[12] but it is noticeable that new build became larger and began to rely heavily on scale, massing and an imposing portico to achieve a feeling of monumentality (SAVE 2004). An early example of this shift is Maidstone sessions house (1810–1819), which is described by Chalklin as the most ambitious to be undertaken in England before the late 1820s (Chalklin 1984). When Robert Smirke took on the commission for the new courthouse at Maidstone he was able to boast that: 'There are no courthouses larger in dimensions than this would be when erected, nor do I believe there is more than one of equal size' (Smirke 1824).[13] Despite this, it is noticeable that the achievements of Maidstone were dwarfed repeatedly in the decades which followed as courthouses became more and more clearly associated with a particular aesthetic that aimed to reinforce the idea that they were unusual or special places. This is reflected in architectural reviews of the time which commonly described courthouses as majestic, dignified, sombre, splendid, elegant, magnificent, monumentally grand, exhilarating, enthralling, foreboding, pompous, heavy, grand and dignified (Mulcahy 2007).

The field was one in which provincial architects, or those with substantial provincial practices appear to have excelled during this period (Watkins 1982; Jenkins 1963). It is certainly the case that architects such as Robert Smirke, Thomas Harrison of Chester and John Carr played a leading role in adjusting the grand designs of classical antiquity to building types such as the law court. Notable examples of courthouses built in this era include Harrison's Chester Shire

The heyday of court design? 117

Hall (1788) described by Henry Colvin and Pevsner (1976) as part of the finest group of Greek revival buildings in Britain (SAVE 2004); Durham County court (1809) and Devizes Assize court (1835) with their Tuscan and ionic porticoes; the stately Salisbury Hall (1788–95); the Italianate Reading Assize courts (1860–1) as well as the neo-gothic Bristol Assize courts (1865) and Bedford Shire Hall (1878–81). It is also clear that this new taste for grandeur was not limited to the Assize courts. The massive Doric columns of Spilsby Sessions House (1824–6), 'the lushly classical' (Graham 2004: 377) or 'lavish' Liverpool County Sessions House (1882–4), and 'sumptuous' Birkenhead Sessions House (1885–7) (SAVE 2004: 63) could well have served as appropriate venues for superior courts. Even the range of police courthouses built across London to the designs of John Dixon Butler between 1905 and 1914 introduce a majesty to a lower jurisdiction which would have been largely unheard of in the eighteenth century.[14] The fact that 25 per cent of courts in use today are housed in Victorian buildings which are of sufficient architectural significance to have been 'listed' could be taken as another sign of architectural achievements during the period (SAVE 2004).

These trends towards architectural sophistication were encouraged in part by the increased availability of new architectural manuals which drew on classical Greek and Roman designs and the growth in the popularity of the Grand Tour.[15] But a general resurgence of interest in the classical form is also reflected in court architecture during the era. The court designs of John Soane, Thomas Harrison and Robert Smirke were at the forefront of the introduction of classical design to England. Indeed, SAVE (2004) have argued that if there was a Sessions House style by the early nineteenth century it was typified by the classical temple style, pedimented and porticoed building with lower wings flanking the central court-room.[16] Courthouses adopting this style reflect more than a blind acceptance of popular neo-classical templates. It was the ideals with which the style was associated which also explains their popularity. McNamara (2004) argues that in addition to the attraction of the Greek neo-classical as a perfect mathematical and symmetrical form it also brought with it particular associations with the birthplace of democracy. In a similar vein the attraction of Roman classicism is attributed to its romantic association with an active and serious civic sphere.[17] Girouard (1990) argues that from the eighteenth century there was a growing acceptance that a sober and solemn Doric order or chaste Ionic was more suited to courthouse design than the more frivolous Corinthian.

The classical template for Assize courts did not go unchallenged. Like all building types of the time, the law court was subject to intense debate surrounding the 'battle of the styles' which led to the supremacy of the neo-classical being challenged by gothic revivalists.[18] This debate about styles raged from the late eighteenth century well into the nineteenth but appears to have had relatively little impact on the design of Assize courts. Just twelve of the 41 courts built to house the Assizes in the nineteenth century were gothic although the style does feature much more prominently from the 1850s, a factor which can be explained in part by the extensive praise surrounding the new Houses of Parliament (1840–52)

Figure 6.1 Manchester Assize Courts (1859–65) © Manchester Archives and Local Studies.

built in a gothic style to the designs of Charles Barry. Alfred Waterhouse's highly influential Manchester Assize courts (1859–65), shown in Figure 6.1, also went a long way to establishing the suitability of the gothic to justice facilities but it was debate about the most appropriate style to be adopted for the new Royal Courts of Justice which really brought the issues to the fore in the years that followed. While for some the neo-classical building could be read as a monument to a healthy public sphere, others felt that Italian and Grecian-inspired edifices were an anachronism in attempts to represent the values held dear by English law. By way of contrast, having outlived its associations with frivolity and ecclesiasticism, the gothic came to be viewed by many as an essentially English style with a romantic link with a medieval and pre-industrial England[19] well suited to represent the antique majesty of the law (Port 1968). Such claims were no doubt attractive to common law lawyers keen to reinforce the idea that the sources of this branch of the law went back to time immemorial.[20]

Temples to justice

A number of courthouses of the nineteenth century stand out as particularly bold examples of the monumentality of the buildings being constructed to house courts. The most obvious place to start is with the Royal Courts of Justice as for many commentators the Courts in the Strand remain the most iconic courthouse of the nineteenth century. In his seminal account of the history and politics of the design process Brownlee (1984) refers to the building as the greatest public building of the generation with a symbolic importance that nearly equals the Houses of Parliament. The award of the commission to design the court, which was generally considered to be one of the greatest architectural prizes of the era, provided George Edmund Street,[21] with a much longed-for opportunity to see his formula for monumental but secular gothic architecture realised in the heart of London (Brownlee 1983). The building was a legal and artistic milestone which was more lavishly decorated than any other court which had gone before it. The Courts stand on a 6-acre site which occupies the whole block between the Strand and Carey Street and considerable resources were expended on the building. The site cost Parliament £1,453,000, the building £700,000 and the oak, work fittings and decoration a further £70,000. The cathedral-like central hall, which forms the main architectural feature of the interior, has been described by Pevsner (1976) as being of considerable architectural merit.[22] Those who worked in the Courts will also have found much more extensive facilities than those available at Westminster and Lincoln's Inn from whence they moved. Many of the facilities for use by lawyers such as legal libraries and dining rooms, which Soane had been forced to remove from his earlier plans for replacement courts at Westminster, were now found in the new courts in abundance. The building accommodated 19 courts and upwards of 700 rooms and offices including a range of waiting rooms, refreshments rooms and libraries.

Despite the significance of the Royal Courts and the attention lavished on them it could be argued that there are better specimens of symbolic courts built to house the Assizes in the provinces. It is certainly the case that there are much earlier precedents for unusually grand law courts. Whilst one would expect most expense to be lavished on the most important courts in the capital city, it has been argued more recently that it is the appearance of provincial temples to justice where previously there had been little of architectural merit, let alone notable courthouses, that is one of the architectural phenomena of the era (Graham 2004). The location of these courts in cities which had prospered from the economic revolution is also no coincidence.[23] Their design and the particular messages they sought to convey appear to be closely bound up with narratives around the new social contract which evolved during this period. The first courthouse worthy of discussion in this context is St George's Hall in Liverpool (1841–7) which was completed 35 years before the Royal Courts of Justice to house two concert halls and Assize courts and is shown in Figure 6.2.[24] Every aspect of design of the building seemed to set new standards. It was described by Pevsner (1976) as the finest and most original classical building of the period and by Sharples (2004) as having one of the greatest interiors of any Victorian building in England. Together with the Walker Art gallery, the Picton Reading room and numerous sculptures erected in the vicinity of the hall[25] it forms a major public space which reflected aspirations on the part of local politicians to challenge perceptions of Liverpool as an uncultured place. As Sharples (2004) has suggested:

Figure 6.2 St George's Hall in Liverpool (1841–7) © Linda Mulcahy.

Size was important if St George's Hall was to express adequately the pride and confidence of the thriving town ... The inscription over the portico, *Artibus, Legibus, Consiliis* (To Arts, Laws and Counsels) proclaimed unequivocally that this huge edifice was a monument to civilised values, in a town where the previous largest buildings had been dedicated to commerce. (p.50)

Everything about the building was ambitious. This cathedral-sized building was designed by Harvey Lonsdale Elmes[26] at the tender age of 25 and was funded in part by public subscription. The great concert hall is an impressive 169 feet long and 74 feet wide and has the largest barrel vaulted ceiling in the United Kingdom and a Minton tiled floor which includes 300,000 hand crafted tiles. The 'small' concert hall on the level above can seat 1,200 people and accommodate an orchestra of 60 and chorus of 70. The building was the first public facility in the country to incorporate an air conditioning system, and the musical organ in the great hall was the largest in the nation when installed.[27] Sharples (2004) reports that size was so important to those commissioning the building that Elmes was asked to draw a cross section of the building which could be superimposed on plans of other great buildings such as Westminster Hall, St Paul's Cathedral and the new

The heyday of court design? 121

Birmingham Town Hall with a view to surpassing all of them. Two hundred and ninety men laboured on the site at its busiest and at a total expense of £290,000 the building cost just over 40 per cent of the magisterial Albert docks which had provided the foundation for the industrial wealth enjoyed in Liverpool.[28] The two Assize courts within St George's are as lavish as their surroundings with their high roofs, wood-panelled interiors and majestic red and grey granite columns supporting the roof.[29] When the courts were sitting the great concert hall was designed to act as a *salle de pas perdu* for litigants. Elmes' intention was that each of the courts should be entered from the great hall through sets of highly elaborate and columned doors. The architect attached great importance to the judges from each court being able to see each other across the great hall when the two doors were open. This sightline, to which Elmes apparently attached great importance, was intended to evoke no less than memories of the vistas in Blouet's restoration of the Baths of Caracalla (Sharples 2004).[30]

Cuthbert Broderick's Town Hall in Leeds (1853–8), built just a few years after St George's Hall to house the local government offices, a large public meeting hall, and four courtrooms also set out to challenge previous standards. The estimate for the building was £45 thousand in 1851 but was soon exceeded as local dignitaries were attracted to the idea of building a municipal palace to rival the best town halls on the continent. These ambitions were fuelled by representatives of the Leeds Improvement Society and others who pleaded for the abandonment of a utilitarian attitude to the project[31] with the result that the final cost of the building amounted to £111,739.[32] Commentators on the success of the project have focused much attention on the great basilican hall around which all the other facilities are arranged. At 92 feet high, 161 feet long and 72 feet wide it is capable of seating 8,000 people. Its elaborate plasterwork and decoration, stained glass windows and sculptures have earned the hall an international reputation. As at Liverpool, the law courts occupied a dominant position in the building flanking the great hall. Linstrum (1999) has argued that the civil court in the north east corner of the building, with its pilasters and curving cast iron-fronted gallery, was particularly impressive. Indeed, the great legal reformer Lord Brougham was later to claim that the new courtrooms within the town hall were unequalled in their arrangement (Briggs 1993).

As with St George's Hall the success of the hall and its attractions were measured in terms of size. This has led Asa Briggs (1993) to conclude that although the inhabitants of Leeds might not have been able to argue long on the merits of the renaissance style adopted by the architect they could appreciate the statistical tables which related to it. Size appears to have provided a useful substitute for other measures of success in the local newspapers which regularly compared the dimensions of the public hall with others such as the Free Trade Hall in Manchester, Birmingham Town Hall, St George's in Liverpool and Westminster Hall. It was with considerable satisfaction that they concluded that the town hall was only inferior in size to the latter two.[33] It was also proudly reported that the musical organ was so large that staff in the factory where it was made were able to hold a

Figure 6.3 Leeds Town Hall (1853–8) © Leeds City Council.

dinner party in the swell box (Grahame 2003).[34] Such were the accolades directed at the building that Leeds Town Hall began to be seen as the model town hall of the time and was much copied across the world. Others have designated it one of the most famous examples of Victorian civic architecture anywhere (Briggs 1993; Linstrum 1999).

Both St George's Hall in Liverpool and the Leeds Town Hall were late examples of the shared building type. It was left to Manchester Assize court (1859–65) to realise the full potential of the sole use and purpose-built courthouse. The opportunities provided by a building dedicated only to law were seized upon with enthusiasm by Waterhouse whose uncle and brother were both lawyers. Although considered an architectural success it was the sophistication of the internal planning that really marks the court out as exceptional.[35] Such is the significance of Waterhouse's design that it could be argued that it was not until its completion in the mid nineteenth century that the modern courthouse was born (see further Mulcahy 2008b). Viewing Manchester Assize courts soon became a prerequisite for anyone involved in court design[36] and on the death of the architect, *The Times* reflected that Waterhouse's design set a new model which was frequently followed in subsequent buildings of the same class (*The Times* 1905; The *Derby Mercury* 1864).[37] As Goodhart-Rendel (1963) has argued:

The heyday of court design? 123

The philosophy of planning is not what the common man can evolve for himself, and although in some architects, particularly Sir Charles Barry and Waterhouse, it seems to have been instinctive, most had to learn it, which in Victorian England they had very few opportunities of doing. Few among the remarkable public buildings to which the Manchester Assize courts led the way are planned with anything approaching the skill displayed in their forerunner (p.97).

The courts were built following a national architectural competition which attracted considerable interest in the press and amongst the public. As *The Builder* (1859) explained:

The exhibition of designs for the Assize courts in Manchester, whether from the labour and cost expended on the drawings, or the merit to be found in the works submitted in the competition, is assuredly, speaking professionally and in the interests of art, one of the remarkable occurrences of our time; indeed in some respects it is not less important than the only exhibition with which it can be compared, that of the designs for the government offices (p.289).

The courts were demolished during the Blitz in 1941 but contemporary accounts and the many architectural drawings available at the Victoria and Albert Museum make it clear that the court was built to impress. At a cost of £120,000 the building contained a combination of courts, offices, consultation rooms, libraries, dining rooms, robing rooms, press mess, waiting rooms, secure underground cells and judges' lodgings.[38] Three stories in height, it occupied a central location near the railway station and together with the adjoining judges' lodgings it took up 350 feet of frontage on Ducie Street and measured 200 feet from its base to its highest point.[39] The central hall, which the visitor encountered as soon as they had descended the three flights of stairs from street level, had dimensions of 100 feet by 50 feet and was 70 feet high, with an ornamental roof and high windows.[40] Decorated with statues depicting the Virtues, famous lawgivers,[41] representations of eight ancient forms of punishment, and local dignitaries, the building was extensively praised for the level of detail.

The success of the Manchester courts is reflected in the fact that Waterhouse was asked to provide preliminary designs and judge the architectural competition for the final monumental court to which I want to allude here, the Victoria Law Courts in Birmingham (1885–6). The stylistically eclectic design of Aston Webb and Ingress Bell which won the competition for this court complex was hailed as representing the spirit of modern architectural design and a worthy representative of the best of English architecture (*Birmingham Pictorial* 1891a; *The Penny Paper and Illustrated Times*, 1891). The £113,000 building housed courts for the Assizes, Quarter Sessions, Petty Sessions and coroner's courts under one roof. The central hall with its stained glass windows and Arts and Crafts-style detail is impressive and the building was considered to be one of the finest pieces of red terracotta work in the country when built.[42] The provision of a large barristers' reading room, robing rooms, separate waiting rooms for male and female witnesses, judges' chambers and library reflect the fact that the sort of facilities mooted by Soane in his Westminster plans and first provided in abundance at Waterhouse's Assize courts at Manchester had now become the norm for any courthouse which aspired to meet the standards of the age.

Symbolic courts and civic pride

The construction of such monumental courthouses in the nineteenth century is commonly understood as just another example of a building programme designed to reflect civic pride in newly emerging cities. In his seminal work on the Victorian city, Asa Briggs (1993) discusses the construction of Leeds Town Hall as one such case study. Motivated by the need to advertise the wealth and success of the growing metropolis, a distrust of the claims of London and local competition with Bradford, he suggests that it is to the new cities of the North and Midlands, rather than London, that future historians will look to find the greatest Victorian buildings of the age. The 1851 census christened England the first industrialised nation[43] and Manchester, Leeds, Birmingham, Liverpool, Bradford, Newcastle, Sheffield, and Glasgow were all at the vanguard of the development of the modern metropolis. The emergence of the phenomena which we have come to call civic pride was, in part, a reaction to the numerous criticisms of conditions in the industrialised North and Midlands. The unnatural conditions of life experienced by the workforce in industrial centres had no precedent and attracted a considerable amount of negative commentary from politicians, journalists, novelists and other writers. Observers were appalled by the ugliness of the new conurbations, the absence of sunlight, and the smell and noise which came with this growth. Briggs (1993) has described Manchester as the first 'shock city' of the nineteenth century and as an environment in which '[t]he "din of machinery" was the music of economic progress'. Overcrowded residential quarters and poor housing caused a crisis in public health and it was almost impossible to control or alleviate outbreaks of typhoid, typhus, smallpox, cholera, rickets, scarlet fever, measles, and whooping cough (Hunt 2005). Significantly, the moral dangers of the new cities were seen to be as great as the physical ones. Depravity, loss of personal control, decreasing religious observance, and the breakdown of sexual mores were extensively commented on. But the working classes were not the only target of criticism. The ethos of the industrial middle classes was seen by many to have brought in its wake a new form of feudalism based on the discipline of the factory. According to this view, the unprecedented civic disarray which was much remarked on by contemporary commentators was blamed on the ethic of self-serving individualism devoid of human emotion which had become associated with industrialisation.

In an era in which it was generally assumed that architecture helped form the character of inhabitants, early Victorian cities were criticised for failing to instil any inspiring vision of community through their buildings. In the initial stages of

The heyday of court design? 125

their growth building often took place without close oversight of urban planning by local government and this had obvious repercussions. Briggs (1993) comments that few of the early buildings in these new cities were distinguished and those that were had generally been blackened by smoke. Midway through the century *The Builder* (1848) remarked on the paucity of the aesthetically pleasing in Manchester in the following terms:

> The Manchester of 1800 and the Manchester of 1848 are the same but in name: within that time its population has increased fourfold, and it has become representative of a new system and new powers . . . The useful has been the prevailing object in Manchester; but until very recently one of the greatest utilities, the beautiful, has received little attention. Science has been thought of more than the arts, and amongst these architecture had not received even a fair share of such small attention as was given to them (p.577).

As advocates and beneficiaries of industrialisation began to enter public debate about the state of the new metropolis, they began to promote alternative images of the industrial city. Rather than focusing on squalor, their emphasis was on the modern city as a symbol of progress in which workers were liberated from the bonds of ancient feudalism and bound only to those they chose to work for. The shift they sought to manoeuvre was from a vision of the city as a cradle of social disorder to that of a cradle of wealth and new, formative and liberating social values (Briggs 1993). In the midst of this debate wealthy industrialists began to react against the slur that the new cities could be labelled spiritual or cultural wastelands. The political sphere increasingly provided a forum in which these new voices could be heard as well as mechanisms for the new urban squire-archy to legitimise their position and sources of wealth. Many of the new industrial towns were presided over by men whose wealth had been earned rather than inherited, and expertise in managing large-scale public projects soon began to be seen as a more important determinant of power than breeding (Garrard 1983). Jenkins (1963) charts how the emergent middle class were particularly good at making their voices heard through the civic building committees.[44] Whilst slum clearance and other public health initiatives improved the life expectancy of those who bore the greatest burden of industrialisation, the creation of new public spaces and buildings allowed for the celebration of urban space and the wealth created by industrialists.

Rather than just being monuments to their success, the buildings commissioned and overseen by the new merchant princes began to celebrate the new political ideology and power base which they sought to promote. Through their architecture industrial cities began to be seen by many contemporary writers as the ultimate symbols of modernity, in which the political, religious, and commercial freedom of individuals was celebrated. Instead of being seen as an anathema to the idea of civilisation, the contribution of commerce began to be viewed as symbiotic with it.[45] There are numerous examples of the way in which the new ideal of a society based on individualism and personal improvement manifested itself. New buildings constructed during this period such as reading rooms and accommodation for literary, natural history, mechanics, statistical, geological and concert halls were all aimed at education and improvement of the mind. Rather than being remembered for their slums, the new Victorian cities began to be lauded for the rich network of clubs and societies which flourished there. It was argued that these new organisations demonstrated a liberal spirit of voluntary participation and the possibility of self improvement which stood in opposition to the imposed social tyranny of the old feudalism.

These accounts of the many monumental buildings constructed during this era as symbols of civic pride are clearly compelling. A host of Victorian buildings continue to dominate the landscape of modern towns and cities and act as a physical reminder of the ambition and arrogance of the age. But it is noticeable that the legal system and its buildings are afforded no special place in these narratives. Quite the contrary. Other than reviews in *The Builder* and a chapter in Waterhouse's biography by Cunningham and Waterhouse (1992), nothing of substance has been written about the much-lauded Manchester Assize court. Similarly Asa Briggs (1993) in his otherwise excellent account of the Victorian city makes no mention of the much-praised Assize court when discussing Manchester. One could be forgiven for not knowing that Leeds Town Hall included four law courts and that one of its purported aims was to attract a grant of Assize. Linstrum's (1999) monograph on Leeds Town Hall mentions the inclusion of the courts but provides no account of their significance. Whilst Briggs (1993) provides a detailed account of the slum clearance which led to the building of Corporation Street he makes no mention of the Victoria law courts, which found their home in this new metropolitan boulevard. In the sections which follow I argue that the law court is worthy of much greater attention than it has been paid in the story of the newly emerging cities of the nineteenth century and that to neglect it is to fundamentally misunderstand the importance of legal reform to the capitalist project and legacy.

The role of law in emerging cities

Despite being subject to the same demands for beauty and comfort, the law court was important to the capitalist project in a way in which the museum or art gallery was not. The Assizes, in particular, conferred a status which went beyond recognition of a new city as a seat of culture; it conferred symbolic status as an instrument of state-sanctioned and centralised power. A grant of Assize and the right to receive the monarch's itinerant justices was a highly desirable goal for those who sought recognition of the successes of the industrialists. Graham (2004) likens the grant as akin to a 'coronation' and it follows from this analogy that the right to hold the Assizes was not easily won. Several applications had often to be made before the right was granted. So, for instance, Birmingham made repeated applications from 1857 onwards and was not successful until 1883. Significantly, architecture played an important role in bids to attract the grant and the standard of accommodation which was available for the most important regional courts could be decisive in the decision-making process. Indeed, all of the courts used as case studies in this chapter were built because of a grant. St George's Hall Liverpool, Manchester Assize Courts and the Victoria Law Courts were all constructed in the years immediately following the acquisition of the right. In an even bolder move, Leeds built their great edifice to civic pride in mere anticipation of the grant. Public speeches made to the Queen at the opening of Leeds Town Hall made clear that local politicians expected that the right would soon been granted to them because of the beautiful building they had constructed to house her courts.[46]

The architecture of the nineteenth-century law court might also be read as reflecting important changes in the part that law and the legal system played in the new social contract which was emerging in the slow movement away from power being vested largely in the aristocracy. This thesis has been specifically applied to the construction of the Royal Courts of Justice in London. Reflecting on the significance of nineteenth-century reform of the legal system Maitland (1883–4) saw the architecture of the two national courts constructed in the nineteenth century as representing the struggle for reform. In doing so he likened Soane's ill-fated Royal Courts at Westminster in which good design was hampered by tradition with Street's new Royal Courts in the Strand which were purpose-built and guided by modern design principles. Brownlee (1984) has argued in a similar vein in his work on the evolution of Street's design of the courts which he saw as a flagship of change:[47]

> For [Street's] generation the law courts possessed a symbolic and real importance nearly equal to that which the Houses of Parliament held for early Victorians, and his building still mirrors the complex vitality of British civilisation in the middle years of Queen Victoria's long reign. Tied to the political and institutional history of the Victorian era, the courts are a monument to the Age of Reform – particularly to legal reform (p.17).

> Reflecting on the significance of the opening of the courts *The Times* was also keen to stress their symbolic importance. The occasion was considered to be '. . . more than the simple opening of an era in the history of our English Justice, that civil institution which of all others in the entire range of the modern world, has had the longest life in the past, whilst its splendid maturity promises it yet an almost incalculable future' (Aslet 1982: 1462).[48]

It is certainly the case that The Supreme Court of Judicature Acts of 1873–5 marked the end of a major and long drawn out reform programme which had been launched in the 1820s and that it was only as change was implemented that the legal system became worthy of celebration for all but the privileged. Blackstone's commentaries (1765–9) had done much to describe the detailed implications of the Act of Settlement (1701) for the legal system and to reinforce the importance of individual liberty as a foundational principle around which legal rights could be understood. Despite this it remains the case that many of the promises for the legal system debated in the eighteenth century were not realised until much later. Expense, delay, ill-considered rules of evidence, uninformed decision making and the poor fit between cause and remedies were rife at the beginning of the nineteenth century. Lack of intellectual merit in the upper reaches of the judiciary and the party connections which continued to be important for appointment to the bench also aroused criticism up until 1875 (Cornish and Clark 1989). Moreover, reform of the notorious problems of the Court of Chancery seemed to have been abandoned by the beginning of the nineteenth century[49] and the number of parallel legal systems continued to pose serious problems for litigants. In addition to the national courts at Westminster, the Assizes, Quarter and Petty Sessions there existed manorial courts, borough courts,[50] courts of pie powder, Verderers and Stannary courts amongst others. The Palatinates of Durham and Lancaster also provided regional substitutes for central courts and a network of ecclesiastical courts dating back to the medieval period continued to exercise a wide jurisdiction over the spiritual and material life of local communities especially in the fields of family law and probate. Whilst some of the jurisdictions enjoyed by these courts were unique there was also considerable overlap which created extensive opportunities for exhausting an adversary. The different jurisdictions and jurisprudence enjoyed by the courts of common law and equity also meant that the same case might be decided differently by two superior courts depending on the route chosen by the litigants.

Distinctive and complex procedures in each system meant that there were many different ways to start an action and different evidential procedures to be understood. Often, this made the hiring of a lawyer essential but the accountability of the profession remained poor and abuse often flourished as a result. In addition, the large number of legal sinecures allowed officers, such as common law clerks and chancery masters, to charge additional fees to file papers and make copies of documents. All of these problems were confounded by an increased demand for adjudication from a growing and complex society and commercial sector, understaffing of the courts and the short judicial year (Brownlee 1984).[51] It needed a significant period of legal reforms throughout the century to make the legal system into a unified and more rational system which at least aspired to meritocracy and equal treatment of all citizens. The taste for reform was also fuelled in part by the concerns of jurisprudes that a strict doctrine of precedent, so important to nineteenth-century common law, could not be established until a regular hierarchy of courts was in place.[52]

The period of reform which came about as a result was extensive and sustained. Writing in 1883 Maitland argued that almost all the re-fashioning which distinguished the Victorian legal system from that of the fifteenth century were those instigated in the previous sixty years and that the age constituted the most eventful era in English law. He concluded that 'Blackstone would have been much more at home in the Court of King's Bench as it was under Edward the Fourth than in the High Court of Justice as it is under Queen Victoria' (p.10). Aroused by the injustices of the legal system, reformers such as Henry Brougham played a major role in reshaping the legal system during Victoria's reign and a raft of changes came into play as a result. The notorious Chancery jurisdiction underwent a series of reforms from the 1830s onwards[53] and significant reforms of local criminal justice also took place over this period as Justices of the Peace slowly had their jurisdiction restricted to adjudication.[54] New systems of courts were also set up to rationalise the legal system. Police courts were established in 1848 to deal with petty criminal cases without a jury and County courts created in 1847 to deal with small-scale commercial disputes. The latter proved to be particularly popular amongst the mercantile classes and reduced the delays and costs of civil litigation considerably (Polden 1997).[55] Indeed, Cunningham has argued that the County court system was so successful in attracting business away from the London courts that it effectively challenged Londoners to put the superior courts in order. By 1860 the *County Court Chronicle* was able to boast that in the previous year the superior courts had handled a mere 86,277 writs of summons compared to the 714,562 plaints entered in the County courts.[56] Calls for the abolition of the distinctions between the courts of common law and equity also had a long history but it is only after the challenge posed by the new County courts that reform began to be seriously contemplated by reformers. By the middle of the century an impressive list of 'fusion' lobbyists was evident and two government commissions were set up to consider the issue in 1850. This led to a partial solution in the form of the Common Law Procedure and Chancery Amendment Act 1852 which removed many of the inconveniences of the divided judiciary and resulted in a cheaper, quicker and more comprehensible system of justice in the national courts. Fifty years of legal reform finally culminated in the Judicature Acts of the 1870s which delivered a fused system of common law and equity, a rational system of linked courts and recognised the Omni-competence of the High courts.

Whilst contemporary debate makes clear that the Royal Courts of Justice in the Strand were intended to celebrate the culmination of these reforms it could be argued that a celebration of law through architecture was influencing design long before the Royal Courts of Justice were commissioned. The courts built at Liverpool, Leeds, Manchester and Birmingham discussed above also spanned this period of legal reform and those involved in the design will have been aware of the new aspirations for the legal system being debated. Indeed for some, law and the legal system during this period came to be characterised as a major cultural and ideological force. Dicey (1914) for instance, claimed that law began to be seen as a positive instrument of social engineering for the first time in the aftermath of the first Reform Act of 1832 and capable of affecting and leading public opinion. It is significant, in this context, that the People's Charter of 1838, engineered by those whose interest had been disregarded in the Reform Bill, expressed a continuing belief in change through the machinery of law. An article in the *Daily News* of 1858 which discussed the opening of Leeds Town Hall and its law courts by Queen Victoria, captures something of the ambition of the age as it relates to the law courts:

> Such a building is to the community to whom it belongs what the domestic hearth is to the family; it is an object of common interest, a centre of

attraction to all classes and grades of society; and it would assuredly be an evil omen if men became so sordid and debased, so engrossed with the factory or the counting house, as to grudge the means required to create and sustain those great civic structures which are symbols of our civic rights as well as ornaments of our commercial cities.

The building of law courts also began to assume a new importance in the popular imagination. Competitions for, and progress of, the new monuments to law discussed in this chapter aroused considerable interest amongst the general public. This is evidenced by the amount of copy in the specialist and general press and the fact that viewing of competition entries for these large-scale commissions also became a popular pastime. When the competition designs for Manchester Assize courts were placed on display at the Royal Institution in Manchester for public view, more than 1,000 visitors came to see them on the first two days (*Building News* 1859) and 50,000 people lined the route to witness the procession of Judges and Magistrates who attended the cathedral and courthouse on its first day of official use (*The Law Times* 1864). In a similar vein, the *Birmingham Daily Post* reported in 1891 that 47,791 members of the public were admitted to see the new courts in Birmingham in a three-day period. At the laying of the foundation stone of Leeds Town Hall a vast crowd of 60,000 accumulated to see the building being dedicated to freedom, peace and trade.

For some, the public buildings which emerged in the nineteenth century also began to challenge the church or cathedral as the prime public meeting place. The new reading rooms, educational institutions and art galleries of the era encouraged new activities based on the ideals of an enlightened secular life. The very size of the monuments to law discussed above challenged the architectural supremacy of the religious building of the 'High' Church. Nowhere is this challenge more evident than in debate surrounding Leeds Town Hall. In the celebratory luncheon organised immediately after the official opening of the building the Bishop of Ripon made clear in this speech that he had been concerned about the opening of so many buildings in the Leeds area which claimed a spiritual association. To his mind the suggestion that the Town Hall was a place in which justice could be administered, scientific enquiry pursued and knowledge advanced and developed were all high purposes best associated with religion (Briggs 1956). What the archbishop understood very well was that the architectural devices developed by ecclesiastical architects were now being put to other uses. Given the dominance of ecclesiastical architecture in earlier centuries it could be argued that it is far from surprising that architects and those commissioning buildings borrowed architectural techniques from religious buildings in order to legitimate the new public sphere which emerged in the wake of industrialisation. But it is also the case that many nuances in architectural and social change are lost if we see the use of size, ornament and height as mere mimicry. The most important point to be grasped here is not just that legal architecture aped religious architecture in this era but that law became a type of religion and its temples

The heyday of court design? 131

sometimes considered more worthy of attention than the churches they sometimes dwarfed.

Fear

There are other, darker tales of the great courthouses of the nineteenth century which need to be told alongside those of civic pride and celebration of reform. Indeed, one of the most fascinating aspects of courthouse design of the era is that so much of it can simultaneously be read as inspiring the celebration of law and as evidence of its violence. What may appear as an uplifting monument to law by those who made their living by it or benefited from its doctrines can as easily be conceived of as repressive and degrading by the classes of society more likely to appear in the dock. An alternative understanding of these nineteenth-century buildings is as crude assertions of authority over a working class conceived of as threatening to the social and economic security of the new social order. In the aftermath of the French Revolution, the demise of feudalism and the rural economy, the birth of Chartism, and the rise of trade unionism, the Establishment clearly had much to fear from the new urban working class and immigrants who flocked into the new industrial towns for work. Indeed, what emerges in the public debate of the time is an increasing association of the working class with space that was dangerous, ugly, dingy, and disease ridden. In this context architecture was constantly employed to segregate the middle classes from the poor.[57] The symbolic courts of the nineteenth century being discussed here did not just seek to erase negative images of the industrial cities from the minds of critical social commentators. In some cases they literally eradicated the unsanitary, or moved it elsewhere; both the Royal Courts of Justice and the Victoria Law Courts were built on the site of former slums causing *The Builder* (1882) to remark in relation to the former that the Majesty of Justice would soon occupy her new palace and law would reign over the spot where lawlessness had recently been rampant.

Fear of unrest was far from unfounded as the communal solidarity of an arcadian agricultural economy was replaced by the divisiveness of industrial capitalism. Boom years in the early 1820s led in turn to a slump and the break out of discontent by the end of the decade. A number of disturbances broke out in the course of the Reform Bill struggles of 1831 and 1832 and the fear of revolution was exacerbated by the economic depression in 1836 which brought with it large-scale unemployment. Cycles of falling wages and unemployment made all too clear that the trickle down effect of industrialised wealth was not as inevitable as early theorists of classical liberalism and political economy had imagined (Hobsbawm 1962). The result was that for many the self-contained privacy and 'empowering' lack of hierarchy in the modern city remained unfulfilled (Hunt 2005). It might be imagined that it would take more than the provision of law courts, parks, reading rooms and museums to convince the populace that industrialisation had somehow liberated them. For much of the nineteenth century the local and central government system was undemocratic, largely unaccountable and often served to protect the interests of those who held office rather than in the public interest. With no effective machinery of modern government, a disturbed social system which lacked the benevolent influence of natural gradations, and an unpredictable economy, many occupants of the modern city posed an ongoing threat to the idea of social order on which statesmen and moralists of the time loved to dwell. Indeed, amongst many quarters there was constant concern that the new conurbations could no longer be 'controlled' (Briggs 1993).

Fears of civil unrest and distaste for mingling with the working classes must have made expenditure on law courts particularly attractive to the ruling elites keen to encourage a fear of the law and penalties it could impose on those involved in campaigns aimed at upsetting the new social order. Law courts built during this era were not accessible and were not intended to be. As I have mentioned in previous chapters the great halls of these monuments to law which have been so praised by architectural historians and lawyers for their beauty were often not open to everyone and spectators and ordinary litigants were increasingly herded around the margins of the new law courts and constrained within enclosures once they arrived in the courtroom. It was in the interiors of Victorian courthouses that traditional features of the court first became highly exaggerated and began to convey a striking sense of place and hierarchy. By way of example, the 'wedding cake' design for court interiors with its strategic use of height was highly effective in promoting a particular organisation of people in court. Viewed in this way it could be argued that courts did not glorify the individual litigant so much as educate the public to be fearful of law. As the *Manchester Guardian* concluded in an article which discussed the opening ceremony for the Manchester Assize courts, 'there is no better school for the study of the spirit of English institutions than that which is afforded to an intelligent and thoughtful spectator by one of the superior courts of law' (*Manchester Guardian* 1864a: 4). The representations of ancient forms of punishment which adorned the public area of the building, such as torture by pouring water down the throat or by weights, conveyed rather cruder messages (*Manchester Guardian* 1864b).

The monumental courts of the nineteenth century can just as easily be seen as monuments to trade and the achievements of the middle classes as they can to law and an independent legal system. The vast upsurge of productivity which came with industrialisation swelled the middling ranks to a size and proportion that compelled a political reckoning. In part this was played out in public debate about suffrage but there were other much more subtle ways in which the economic power of the new middle classes could be asserted. Law and its building projects provided an attractive and early outlet for the new elite. After 1832 the judiciary increasingly came from middle class whilst the polite professions of church, armed forces and civil service remained in the preserves of the old upper ranks for much longer. The office of Justices of the Peace which had always been the prime institution of the ruling class also began to be infiltrated by the new middle classes. As the new manufacturing districts developed, the mill and mine owners were drawn on to the bench to continue the hostilities over discipline and conditions that infected many workplaces. Moreover, McNamara has argued that the trend

The heyday of court design? 133

towards increasingly grand buildings for the courts was also encouraged by middle class lawyers and architects as a means of securing public affirmation of their still fragile status (McNamara 2004).

The link between law and the market place was commonly made very explicit in the fabric of the law courts during this era. So, for instance, the artwork which adorns the outside of St George's Hall in Liverpool (1841–54) has a series of relief sculptures to one side of the main entrance which celebrate *Justitia* and a series of sculptures on the other which celebrate the great commercial successes and prosperity of Liverpool. Similarly, the archway above the main entrance of Leeds Town Hall shows figures representing local industry, poetry and fine arts, sciences and law. Panels on each side of the doorway show a child holding a fleece above the emblems of power and justice (Grahame 2003). Whilst these sculptures could be seen as a glorification of labour the link between commerce and justice takes on a different hue when one considers that it was in the course of the nineteenth century that legal doctrine served to protect the rights of the industrialist and became hostile to the demands of the worker. There is some evidence of judicial conservatism when it came to protecting the rights of the industrialists or entrepreneurs (Kostal 1994; Rubin 1984) and doubts have been cast on the willingness of the legislature and judiciary to always protect the interests of the business community but it is clear that a host of nineteenth-century laws and common law doctrines often sought to favour the industrialists and burgeoning middle classes (Ferguson 1984). With the development of a free-enterprise economy in the nineteenth century commercial law, in particular, became the indispensable instrument of the entrepreneur and provided a legal foundation for capitalism to flourish. The full recognition of the executory contract during this era by which the parties were bound to future exchanges is closely associated with the commercial or economic changes of the time and reflects the demands for greater protection for careful planning made by the business community (Ferguson 1984).[58]

The new County courts, heralded as the poor man's court, rather ironically facilitated an improved system of debt recovery through the courts which suited the businesses of the middle classes much more than the working class poor forced to rely on credit.[59] Viewed in this way the courts can be seen as an integral feature of a capitalist economy dependent, in part, on the distribution of household commodities and consumer credit. Rubin (1984) argues that it is a much-repeated myth that imprisonment for debt was abolished in 1869 when it was in fact retained in the case of small consumer debt. Despite the many reports which had drawn attention to the extent of poverty in the nineteenth century, lack of concern about the underlying causes of poverty by Parliament, the judiciary and registrars was prompted in part by a widely held view that the working classes were thriftless, extravagant and lacked the will to resist the tallyman or lure of the public house. In Rubin's (1984) words:

> That imprisonment for debt had to await the later twentieth century before abolition took place, is a stunning testimony to the combined strength, over a

134   Legal Architecture: Justice, due process and the place of law

number of generations, of local business and legal elites who patronised most of the county courts. For the insolvent poor, it was their unfortunate fate that the majority of judges, certainly before 1914, remained both resolutely imbued with the spirit of deterrence and also blind to the reality of poverty (p.290).

Other measures which favoured the middle classes included the law relating to divorce, bankruptcy, employment law, partnership, commercial transactions and strike activity and have led Tigar and Levy (1977) to characterise the period as one in which there was a triumph of bourgeois jurisprudence.

Conclusion

This chapter has sought to understand why it was in the nineteenth century that the buildings which housed courts reached a peak of 'architectural elaboration' (Graham 2003: 267) which has been largely unmatched since. Traditionally accounts of the period by architectural historians have treated law courts as just one example amongst many of the building revolution which emerged in the nineteenth century. Where they have acknowledged a judicial function for these buildings social historians have understood the courthouse as a mere example of civic pride. I have attempted to argue here that in failing to look more closely at the role of the Victorian law court historians have underestimated its particular importance in the long nineteenth century. It could be claimed that the reformed legal system of the era was one which was more rational, accessible and responsive than those which had gone before it. The secular courts gained more authority than they had previously as much of the jurisdiction of the church courts was taken away from them, the fusion of law and equity reduced complexity and the birth of county courts and police courts improved the efficiency of debt collection. Not only were the law courts overseen, funded and designed by the middle classes but their size and ambience can also be interpreted as glorifying the achievements of those with newly acquired and negotiated capital and economic strength. Architects of the law courts discussed here may have aped the styles of the past and debated long and hard about the relative merits of the neo-gothic or classical style but the shrouding of the court in ancient garb masked the many ways in which they housed a very new type of legal system. In the next chapter of this book I go on to consider whether the ambition of contemporary architects matches those of their nineteenth-century forebears and what the current state of courthouse design tells us about the role of law in modern-day society.

Notes

1 The term 'long nineteenth century' was defined by Eric Hobsbawm as the period between the years 1789 and 1914. See further Hobsbawm (1962, 1975).

The heyday of court design?   135

2 SAVE (2004) have argued that Northampton Sessions House was one of the most architecturally sophisticated courthouses of the century.

3 In the case of Assize courts legislation passed in 1769 clarified that quarter sessions had the power to repair shire halls.

4 The poor standards at Westminster prompted the government to commission John Soane to build new accommodation for the courts between John Vardy's stone building of 1755–99 and the medieval Westminster Hall. Soane's original plans show considerable merit but interference from Parliament, arguments about the style to be adopted and the restricted site meant that that building was almost doomed from the start. Having started construction according to his Palladian design they were demolished at the request of Parliament and the architect was required to re-design in the gothic style.

5 He argued that Stafford placed too much emphasis on the county rooms to the detriment of the courts; York was too small; and Lancaster and Chester had poor acoustics.

6 There was extensive debate about the location of the courts with much support for the idea that they should be located in Lincoln's Inn (see further *The Builder* 1854; *The Law Times* 1859). Locating them on the Embankment was also mooted. For a full account see Brownlee (1984).

7 So for instance, Robert Smirke who had the largest architectural practice of his day in London worked on a higher proportion of public buildings than previous generations of architects. In the course of his career he designed over 50 public buildings, 60 private houses and 20 churches (Riddell 2004; Summerson 1986b; Crook 1967b).

8 By the time that the foundation stone of Maidstone Sessions House was laid Smirke had already designed six buildings to house courts and he went on to design Shrewsbury Shire Hall (1833–7).

9 From 1870–1914 a number of public bodies stamped the London suburbs with a recognisable, repetitive building type that contributed to the capital and its suburbs' identity. These included school boards, the post office, the Fire Brigade Branch of the London County Council and the Metropolitan Police Authority. Each of them established permanent design and planning teams whose designs were influential as they were often borrowed by boroughs across the country (Graham 2003).

10 Chalklin (1998) cites the example of Baron Hotham who laid a fine of £2,000 on Shropshire in order to compel the local justices to provide a new Shire hall for the Assizes at the summer Assize of 1782.

11 For a detailed review of the cost and incidence of new build over this period see Chalklin (1998).

12 See for instance the accounts of Victorian courts in Lewis (1982).

13 The Sessions House was one of the largest and most expensive of the era. Of the 18 county courthouses built between 1800–30 Maidstone court is listed as the third most expensive after those in Carlisle and Gloucester (Chalklin 1984). At a final cost of £29,106 it was well above the £20,929 average.

14 These are Great Marlborough Street Magistrates' court, West London Magistrates' court, Tower Bridge Magistrates' court, Old Street magistrates' court and Clerkenwell Magistrates' court. SAVE (2004) describes them as having a sophisticated free classicism, elegant pediments and distinctive interiors.

15 Andrea Palladio's *The Four Books of Architecture* was first published in four volumes in 1570 and contained a number of engravings by the author. The first English edition was published in the eighteenth century. Colen Campbell's *Vitruvius Britannicus*, produced in three volumes between 1715 and 1725, was one of the first architectural works to originate in England.

16 In addition to the courts at York and Liverpool mentioned above, fine examples of the neo-classical courthouses of the era include Beverley Sessions House (1805), Newcastle Moot Hall (1810), Knutsford Sessions House (1817), Guildford Cornmarket (1818),

Ely Shire Hall (1821), Worcester Shire Hall (1834) and Robert Smirke's neo-classical courts at Hereford (1815), Gloucester Shire Hall (1815), and Shrewsbury Shire Hall (1833–7).

17 Whilst architectural purists argued that the classical form should only be used in buildings familiar to the ancient Greeks and Romans, others such as Robert Smirke argued that it could and should be adapted to new building types such as the standalone law court.

18 The style of buildings in the vicinity might often dictate design. So for instance Smirke chose a gothic style for the courts at Lincoln and Carlisle because of the proximity of the medieval castle and a Tudor citadel. Harrison came to the same decision at Lancaster Castle when building the Shire Hall (1786–99) (Watkins 1982, Jenkins 1963).

19 Gothic architecture originated in twelfth-century France and when introduced to England was originally known as the French style. Like gothic architecture across Europe it is defined by features such as pointed arches, vaulted roofs, buttresses, large windows made possible by new forms of weight distribution and spires. In time England developed its own form of gothic architecture which laid emphasis on horizontal lines. I am told that the English gained a reputation for particularly spectacular gothic ceilings.

20 In the event all 11 of the invited designs submitted for the Royal Courts of Justice were in the gothic style, although the neo-classical was not without its advocates (see further Brownlee 1984). The law court was to play another significant role in the story of the battle of the styles for the Royal Courts of Justice was the last great public building to be commissioned in this style and are now remembered as its swansong (Port 1968).

21 In common with many other national architectural competitions in the era there was a considerable amount of controversy surrounding the appointment of the architect. The commission was originally offered to George Edmund Street and E.M. Barry jointly, a suggestion which neither of the architects favoured. See further Port (1968), Brownlee (1983).

22 The hall is 238 feet long, 48 feet wide and 80 feet high (Royal Courts of Justice 2005).

23 Manchester was given city status in 1853, Liverpool in 1880, Birmingham in 1889 and Leeds in 1893.

24 Originally it was envisaged that separate buildings to house the Assize courts and concert halls would be constructed and two competitions were held to identify an architect. Harvey Lonsdale Elmes won them both. In time Elmes' suggestion that the two should be combined in an exceptionally large building was accepted (Sharples 2004). It has been claimed that it is possibly the only building where you could be tried for murder, have a ball or listen to a concert all under one roof: see online. Available HTTP <http://www.bbc.co.uk/liverpool/culture/2002/08/st_georges/map.shtml> (accessed February 2009).

25 For a detailed account of the sculptures see online. Available HTTP: <http://myweb.tiscali.co.uk/speel/place/lpoolstgeo.htm> (accessed January 2009).

26 Elmes died of consumption in Jamaica in 1847 where he had gone to avoid the English winter and in 1851 C.R. Cockerell who had already been acting as an adviser from 1849 was appointed the chief architect. Cockerell altered the interior design quite radically and the decoration of the interiors is largely due to him (Sharples 2004).

27 It enjoyed this accolade until a larger one was installed in the Royal Albert Hall in 1871. See further online. Available HTTP: <http://www.stgeorgeshall.eu/building.html> (accessed August 2009).

28 The Albert docks were built by Jesse Hartley in 1845 and set the standard for many later docks.

29 The building, which has recently been restored, is open to the public and visitors can go into the Crown court. It can also be seen in the film *In the Name of the Father* directed by Jim Sheridan.

The heyday of court design? 137

30 When he took over the design of the building on Elmes' death, Cockerell scuppered this scheme when he allowed the musical organ to be placed in front of the entrance to one of the courts.

31 The Leeds Improvement Society was founded in 1851 to suggest and promote architectural and other public monuments in the town.

32 The grant of Assize was given after the Town Hall had been completed. Alterations to accommodate the Assizes in 1864 cost an additional £10,261 (Grahame 2003).

33 See, for instance, *The Leeds Mercury* 1854.

34 The organ was 50 feet high, 47 feet wide and 27 feet deep. It weighed almost 70 tons (Grahame 2003). See further *The Leeds Mercury*, 1858.

35 Waterhouse's approach to the gothic style was considered rather eclectic and not always approved of by purists.

36 For instance William Gladstone, one of the judges for the Royal Courts of Justice competition visited in October 1864 (Brownlee 1984). In their preparation for the building of Victoria Law Courts, the mayor, town clerk, and alderman toured courts at Manchester, Leicester, Chester, Shrewsbury, and Worcester.

37 See for example. Liverpool Sessions House (1882–4), the Victoria law courts at Birmingham (1887–91), and alteration at Winchester Great Hall and Law Courts (1871–4). The design for the courts at Manchester were explicitly praised in the instructions for the competition to design the Royal Courts of Justice (see further Brownlee 1984; Waterhouse 1864–5; Cunningham and Waterhouse 1992).

38 These were built to the side of the court complex but connected with it.

39 With dimensions of 59 feet by 45 feet the courts and public galleries were large when compared to other Assize courts and Waterhouse described them as being amongst the largest in the Kingdom.

40 See further the extensive description *The Manchester Times* (1859).

41 These were Alfred the Great, Edward I, Henry II, Ranulf de Glanville, Gascoigne, Sir Thomas More, Bacon, and Sir Matthew Hale. Fortunately, three of these statues (Sir William Gascoigne, Sir Thomas More, and Matthew Hall) survived bombing in the Blitz, together with one of Mercy, and until recently were displayed in the waiting hall outside the Courts of Justice in Crown Square, Manchester. They are remembered in a collage in the judicial dining room of the new Manchester Civil Justice Centre.

42 In due course it became so influential locally that by 1897 Birmingham had become renowned for its great terracotta buildings.

43 The census revealed that over 50 per cent of England's population was resident in towns and cities.

44 The results were not to everyone's taste, and for many decades it was fashionable for architectural historians to condemn the period as a failure in architectural terms. In this vein Hobsbawm (1962) offers the following analysis of the new industrial class: 'Their wives began to turn into "ladies", instructed by the handbooks of etiquette which multiply about this period, their chapels began to be rebuilt in ample and expensive styles and they even began to celebrate their collective glory by constructing those shocking town halls and other civic monstrosities in Gothic and Renaissance imitations, whose exact Napoleonic costs their municipal historians recorded with pride'.

45 Nowhere is this more apparent than in Ford Maddox Brown's chronicling of Manchester's mercantile heritage in Manchester Town Hall.

46 The address to the Queen expressed the statement that '. . . in our architectural plans, we have borne in mind the probability that, at no distant time, civil and criminal justice may be dispensed to an extensive region in this town, the real capitol of West Riding'. A copy of the address was sent to the Town Clerk at York, one of the traditional locations for the Assizes in Yorkshire (Briggs 1993). After the grant of Assize was made in 1864 alterations were made to the hall to fully accommodate the Assizes at an additional cost of £10,261. So for instance more cells were needed in the Bridewell.

138 Legal Architecture: Justice, due process and the place of law

47 Built just 42 years after Soane's courts at Westminster those involved in the design of the courts in the Strand were also keen to learn from the somewhat disastrous planning history of Soane's courts.

48 The courts also represent an artistic milestone as seen as the zenith of High Victorian architecture (Graham 2003; Jenner 2000) although there were substantial revisions to the plans in the interests of practicality (Brownlee 1983).

49 A commission, chaired by Lord Eldon, was appointed by Parliament to look into the problems of the Chancery court in 1816 but appears to have produced a report which was no more than an apology for the abuses of the court (Holdsworth 1903).

50 The Tolzey court in Bristol, the court of passage in Liverpool, the Guildhall court in Norwich and the Salford Hundred court are all examples of borough courts which continued to sit until 1971 (Cornish and Clark 1989).

51 In the common law courts the judicial year consisted of four terms of three weeks.

52 This also relied on an accurate system of reporting. A semi official series of law reports was established in 1865 (Stein 1984). Before the age of print precedents were collected in Yearbooks and from the sixteenth century private law reports appeared (Cornish and Clark 1989).

53 This led to a reorganisation of the bankruptcy jurisdiction and appeals process, the creation of additional judicial offices, the reform and eventual abolition of Masters, the introduction of fixed salaries for court staff, the lowering of court fees and restrictions on the number of office copies required from suitors (Holdsworth 1903).

54 They shed responsibility for prosecutions from 1829–56 when the modern police force was established. Many of their general political and executive functions were also lost when municipal councils were introduced in the boroughs in 1835 although governance of the shires in quarter sessions continued until the creation of elected county councils in 1888. Responsibility for local gaols was also ceded to a national system in 1877.

55 The 491 County courts were grouped into 60 circuits and were expected to sit at least once a month, and once a week in populous districts. Sixty judges were employed to adjudicate cases assisted by 450 clerks and assistant clerks. Moreover the *County Court Chronicle* anticipated that just under 2,000 attorneys and 200 barristers would be using them on a regular basis (*County Court Chronicle* 1846).

56 In 1855 the *County Court Chronicle* reported that the 22 busiest County courts were being presented with between 9,829 (Bury) and 19,291 (Wolverhampton) plaints.

57 Cannadine (1977) suggests that this was particularly noticeable in Manchester, which alongside Oldham and Sheffield was one of the only cities in which affluent houses encircled the city completely, so that the middle classes could easily avoid contact with the poor.

58 Ferguson argues that the relationship between the law and realisation of commercial expectation was not always as close as commentators have suggested. It is, of course, a major theme of existing contract scholarship that the formality of contract law does not always suit the flexibility required in the business sector.

59 In his research Rubin (1984) identifies that tallymen or peripatetic salesmen who offered goods on credit were the most frequent users of the County courts to enforce debts. Other tradesmen regularly represented include grocers, coal merchants, butchers, bakers, boot dealers, doctors and landlords. He claims that the use of the courts by tallymen caused a crisis of legitimacy for some members of the judiciary who disapproved of the practices of these traders.

Chapter 7

Back to the future: Is there such a thing as a just court?

Introduction

This chapter explores the symbols of law and our legal system promoted by contemporary legal architecture. It considers the extent to which the architectural visions of Victorian designers discussed throughout this book have since been revisited or revised. The focus is on the issue of whether there is such a thing as a 'just' court and the ways in which the changing context of democracy have promoted new ideas about the public spaces of law. The potential for architecture to reflect and promote innovative concepts of justice has been clearly recognised elsewhere. The design of the Constitutional Court in South Africa provides one such contemporary example of an attempt to engender a republican conception of transformative constitutionalism through design.[1] Moreover it is not only symbolic national courts which have reflected and precipitated changing notions of justice. It has been argued that the focus on community participation in the design of some Koori courts in Australia and community justice centres in Australia, the UK and the US reflect a focus on therapy rather than incarceration or punishment (Smart and Smart 2009; Kirke 2009). In some jurisdictions commentators have argued that as the ideology underpinning common law systems shifts the designer has become a partner in effecting change rather than just reflecting it (Brawn 2009).

One of the interests which has fuelled this project is the extent to which it is possible to both condition the design and design the conditions of the trial. In the sections which follow I consider the ways in which traditional organisation of space in the courthouse which marginalises and sometimes degrades participants can be subverted in the design process. The first section of the chapter looks to the recent design history of courthouses and whether the challenge of revisiting the fundamentals of courthouse design were met during the extensive court building programme of the 1970s and 1980s. Drawing on assessments of recent build in the UK and visits to courts in five legal jurisdictions the remainder of the chapter presents a number of examples of modern courts which attempt to disrupt the canon in favour of creating spaces which reflect contemporary debates about participatory justice. In contrast to the *Court Standards and Design Guide* (Her

140 Legal Architecture: Justice, due process and the place of law

Majesty's Courts Service 2010) assertion that the design of the interior of English courts can not be improved upon by architects it is asserted that exciting new courthouses are being imagined and in some instances built.

Centralisation of design

A major theme to emerge from the recent history of court architecture is the rapid centralisation of design and oversight of the court estate. Until the 1970s funding of buildings for the Assize was largely the responsibility of the county in which courts were held. As I have made clear in earlier chapters this led to considerable variation in courthouse design. In some areas a rich benefactor might fund the building of an impressive monument to the community and law whilst in others money might be raised on the local rate or raised by public subscription. Under this system some counties found their courts neglected because of lack of funds or political will. Whatever the state of affairs, local planners were able, and often keen, to reflect the characteristics or achievements of the local community in design. Buildings commonly use statues, bas reliefs, coat of arms and inscriptions to celebrate justice alongside local commerce or historic events which had helped form community identity. The fact that the county was responsible for the provision of buildings for the Assizes also symbolised the coming together of the local community and the monarch's appointed representatives to try serious offences in unison. Despite the longevity of such arrangements for the Assizes by the 1960s the system was being questioned as outdated and inefficient and this provided policy makers and designers with an opportunity to promote the merits of centralised planning. The reforms which were to follow led to many Assize towns losing their hard-won status as key centres for the trial of serious cases in the provinces.

By the time the system came to be reviewed it was clear that the flurry of courthouse building in the long nineteenth century described in the last chapter had not been matched in the first half of the twentieth century with the result that many of the courthouses in use dated back to the Victorian era and beyond. Two world wars, a depression and a focus on the provision of new domestic housing had all contributed to a lack of attention to justice facilities within the government and design community. In the meantime the post-war rise in crime put an additional strain on courthouses which were already deemed inadequate for modern needs. When the Beeching Commission (1966–9)[2] was set up to inquire into the system of Assizes and Quarter Sessions it received numerous submissions about the state of courthouses across the country which provided detailed evidence of the failure of existing buildings to meet the need of a modern litigation system. Despite the architectural ambitions of the Victorian era discussed in Chapter six, the Commission concluded that by the latter part of the twentieth century:

> We have seen courts with no waiting rooms, no consulting rooms, no refreshment facilities, and with toilet facilities which were disgustingly insanitary.

Back to the future: Is there such a thing as a just court? 141

> Beneath the courts, some of the accommodation for remanded prisoners is so cramped and primitive that prison officers avoid using the worst of it if they can. Behind the scenes, the judge's retiring room may not be much bigger than a cupboard and may, indeed, serve the charwoman in that capacity when its distinguished occupant is gone. (p.47)

The Beeching Report led to the courts of Quarter Sessions and Assizes being replaced by new 'Crown' courts based in larger towns and cities rather than those centres which had been of historic importance. With little prospect of the poor condition of the building stock being remedied in the short term the Committee recommended a thorough review of the estate and the centralisation of responsibility for court design. Most importantly for present purposes responsibility for the provision and upkeep of Crown and County court accommodation across the country was ceded to the Lord Chancellor's Department. This meant that the centuries-old practice of responsibility for design of Assize facilities being vested in the hands of those responsible for local government came to an end. The thirst for rationalisation which was so characteristic of the era of increasingly 'big' government had an important offshoot for courts design; the birth of the centralised planning guide.

It was bodies involved in the construction of the Magistrates' courts, most notably the Greater London Council, who pioneered the idea of common design specifications from the late 1960s onwards (Brown 1980a).[3] A multi-disciplinary working party set up in 1967 to specify needs and give guidance on the design of Outer London Magistrates' Courts published a report which was unique in the field and became so influential that it began to be used throughout England and Wales to steer new build (Brown 1977). In time civil servants, architects and the judiciary were encouraged to translate their collective experience of, and presumptions about, court architecture into a universal template to guide all courthouse designers. Whilst recognising that the law is a 'field well known for strong individualism' and 'strong local judicial tradition' (Home Office and Greater London Council 1977, p.5) the production of the *Guide* was a deliberate attempt to provide a standardised and rational base for the rapid compilation of briefs which gave a clear understanding of the design issues to be addressed (Brown 1980a).[4]

The significance of centralised planning was the challenge it provided for policy makers and designers to reconsider how a late twentieth-century courthouse in a democratic society should look and be experienced. Inactivity in court design in the first half of the twentieth century meant that by the time expectations came to be reviewed there had been seismic shifts in the political and social context in which those debates took place. The introduction of universal suffrage and post-war alterations in the social contract heralded the birth of an era in which public institutions, including the legal system, were expected to be more accountable to the general public than ever before. Early guides reflect a strong sense of the heritage of the legal system and the ways in which architecture had been used to glorify law in the past. But while Victorian ambition continued to inspire architects of the period there was also a recognition that the messages the courthouses Victorians produced conveyed was that the institution of law should not be questioned. There was, for instance, a recognition that courthouses of the past had rarely reflected concern for the comfort of the public or been concerned with images of justice as accessible. A strong sense emerged of the need for a new approach to design which reflected the fact that sovereignty was increasingly seen as vested in the people (Black 2009). In an era in which public institutions became subject to unprecedented levels of scrutiny and criticism there was a sense that architecture had a role to play in mediating and reflecting this new political compact. In the words of Brown (1987) the challenge for court designers was '. . . the resolution of the dialogue between the individual and State – the rulers and the ruled – and a building which symbolizes the health and viability of such a social contract' (p.91).

Debates about the democratic turn were highly visible in the field of architecture. There are many stories of modernism but a unifying characteristic is undoubtedly the ambition of this movement which emerged in the early part of the twentieth century and flourished in the interwar period to renounce the old world (Gold 1997). Its simple undecorated lines, geometric forms and attention to function provided a stark contrast with the neo-classical or gothic form and deliberately drew on ideas of rupture with the past. The utopian ideals of the movement which focused on ideal cities gained much of its strength from the haunting images of poverty, deprivation and alienation associated with the Victorian industrialisation and post-war inner cities. The increasing centralisation of the State and the growth of welfarism simultaneously promised a new society in which inequality could be addressed. Viewed through such fresh perspectives the Assize system with its many historic buildings suggestive of the imposition of law must have seemed antiquated and oppressive and the possibilities offered by a reformed legal system to achieve equality exciting.

The opportunities for experimental design which broke with the traditions of the past were rife and soon realised. Concerns about overcrowding and inadequate segregation of participants in the new Crown courts which had inherited the Assize estate led to an extensive new building programme for courts which lasted from 1972 until 1996 and cost in the region of £500 million. Centralisation of responsibility for the building stock created the opportunity to impose a unified vision of good design on courthouses across the country and the well-funded building programme gave architects and planners the opportunity to see their visions transformed into bricks and mortar. One hundred and thirty nine projects were completed during this period and 382 new criminal courtrooms, a net increase of 212, were built. The building programme also resulted in the creation of a new breed of court, the combined court centre which demonstrated that control of different court systems was finally being vested in unified agencies. Twenty eight of these court complexes, containing up to 20 courtrooms, were created (SAVE 2004). Such was the importance of this initiative that Rock (1993) has referred to it as 'one of the largest programmes of monumental building since

Back to the future: Is there such a thing as a just court?   143

the pyramids' (p.242), seen by some as the last major contribution by central government to Britain's city centres.

Courthouses built during this era were strongly influenced by modernist and emerging postmodernist ideas about design. At an international level buildings such as Le Corbusier's courts at Chandigarh (1950–57) where the majesty of law is represented by massive concrete pylons proved revolutionary in their subversion of traditional design concepts.[5] In an Australian context Katseiris (2009) also names the building of the High Court in Canberra (1975–80) with its use of high ceilings and flat courtrooms as the moment when judicial buildings in the country attained their contemporary voice.[6] In a UK context achievements were more modest but courts such as those built in Plymouth (1961–3) aimed to achieve an appearance of lightness and dignity as a foil to the oppressive mass and solidity of the adjoining Guildhall which had housed the County and Magistrates' courts since being opened in 1874 (Sterling 1963). Courts of this era can also be characterised by less pretentious court interiors which consciously resisted the conventional and superficial trappings of pomp and civic dignity (Brown 1980a). This is particularly true of the interior of courtrooms which became noticeably simpler and flatter, a trend which continues to this day and can be seen in the modern courtroom shown in Figure 7.1. It can be seen from the illustration that there is much less evidence of hierarchy reflected in the vertical dimensions of the room and a conscious rejection of the sort of excessive detailing so loved by Victorian architects. The 'wedding cake'

Figure 7.1 Modern courts are now much flatter than their Victorian counterparts © Her Majesty's Courts Service.

interiors of nineteenth-century courtrooms with their central well and theatre-like qualities were rejected during this period in favour of courtrooms in which additional height was generally only used for the judicial dais and the back row of jurors.[7] Even when used for the dais a convention was developed which limited this raised platform to three steps above the rest of the courtroom.

This is not the place to rehearse the merits of Modernism but considerable doubt has been cast on the success of early experiments in modern court design. There is no doubt that the great achievement of these pioneers was to provide fresh and revolutionary perspectives on court design and a questioning of much-hallowed assumptions about how the relationship between the State and individual should be symbolised in bricks and mortar. But for many critics the new build brought with it excessive utilitarian tendencies which in some cases could be read as brutal rather than humane. The Modernist penchant for simplicity of design was read by critics as austere, an impression which was reinforced when it became clear that many of the new materials being experimented with began to deteriorate more quickly than anticipated. Few interested in the architecture of modern law courts would aspire to the medieval castle as a model but several law courts built in the 1970s and 1980s served to convey the image of a fortified and inward-looking approach to design. Those familiar with the courthouse in Bolton will know that the clear impression of the buildings signalled by the exterior is of a modern-day concrete fortress. Regular users of the buildings also found much to complain about. Many of the first generation of modern courtrooms were criticised by the judiciary and bar for their low ceilings, lack of natural light and natural ventilation.[8] Reflecting on the period, Tony Henocq (1999) of the Department for Constitutional Affairs has concluded that the courtrooms were designed as if they were 'clinical operating theatres with the aim of excluding all external noise. The walls were deliberately featureless and designed to avoid distracting the users and the lighting was mercilessly efficient' (p.7). Such buildings conveyed what another architect has since described as 'a crushing bureaucracy' on courthouse design (Brown 1980: 1196). The success of the architects of the era in breaking with tradition is clear, their success in creating buildings which were accessible or inviting less so.

The contemporary response to this early period of experimentation would appear to be that buildings which emphasise machine-type qualities are now considered inappropriate design templates for courts and are best understood as a product of a particular architectural era (Department for Constitutional Affairs 2004). One reaction to the problems identified in the courts of the 1970s and 1980s would appear to be a more prescriptive approach to the design principles which should steer new build. Whilst the early design guides focused almost completely on technical specifications for courts more recent editions have begun to reflect on the qualities which architects should aim to convey in courthouse design and the tension between the need for an architectural statement and accessibility. In a section which has appeared in design guides since the 1990s it is made clear that law courts should suggest openness:

The Majesty of the Law is a phrase often used as guidance to the designer but there are many ways of interpreting this abstract concept. That the function of a courthouse is fundamentally serious is not to be disregarded. But it must also be remembered that, for most people, attendance in the building will be a rare and perhaps disturbing experience: this consideration should temper the designer's approach. The building should be seen less as a symbol of authority than as an expression of the concept of justice and equality before the law. The scales of justice are a more appropriate symbol than the sword of retribution.

(Her Majesty's Courts Service 2010, p.3)

Elsewhere in the contemporary *Guide* it is asserted that a new court should represent the ideas of equality before the law and that the buildings created should be pleasing to the eye. Perhaps most ambitiously the *Guide* suggests that modern build should 'delight' and 'excite the senses' whilst stretching the imagination (Her Majesty's Courts Service 2010, p.19).

Courts such as Northampton Crown and County court built in the 1990s showed a greater sensitivity to soften the clinical lines of modernist courthouses with its two storey atrium, naturally lit courtrooms, higher ceilings and views of a garden space from the restaurant (*Construction* 1992). The success of buildings such as Southampton Magistrates' Court and Truro Courts of Justice has also been acknowledged in the architectural press. Indeed the latter has been generally praised for the splendour of the public areas, the use of natural light, replacement of dark wooden panelling within the courtrooms with light bright walls and the positioning of consultation rooms overlooking a courtyard (Stansfield 1985). More recently still a number of architects have been keen to revise narratives of a complete break with the past. For architects such as Katsieris (2009) the quasi-religious spirituality of the soaring architecture of the nineteenth century can continue to inspire modernist design. Recent projects in the UK have certainly been completed to architectural acclaim. The new Manchester Civil Justice Centre (2006–2008) is the biggest court complex to be built in the UK since the Royal Courts of Justice and was a finalist in the Prime Minister's Better Public Building Award. Run in conjunction with the 'Government's Better Public Buildings Initiative' (2008) the design team was praised for meeting a demanding brief for 'a sustainable building of civic generosity and European significance with minimal impact on the environment'. The Judges commented that the result was a building which is a 'modern rival to the Royal Courts of Justice' (Better Public Buildings Initiative 2008).

The current state of building stock

Regardless of the many claims of the *Court Standards and Design Guide* (Her Majesty's Courts Service 2010) that courthouses should have a civic presence and that their design's should be associated with high quality build there has been considerable criticism of current building stock. This is nothing new. Even in the

146 Legal Architecture: Justice, due process and the place of law

heyday of courts design described in the last chapter there was regular criticism of courts across the country. What is noticeable about recent commentary is that the state of the building stock is being directly related to concerns about the status of the justice system in the UK and ongoing issues being raised by the academic community about the vanishing trial (Galanter 2004; Resnik and Curtis 2007). Drawing on her considerable experience of undertaking research into civil justice Hazel Genn (2009) concluded in her Hamlyn lectures of 2008 that:

> Having spent time over the past twenty years conducting research in civil courts and tribunals I can confirm the sorry state of the civil courts. When I talk of the crumbling of civil justice I speak as someone who enters the court buildings through the front door with litigants and I walk the public corridors. I have personally witnessed the decline . . . The public areas of some courts are run down and squalid. They resemble the worst to be found in NHS hospitals. But the courts are not outpatients' departments. They are sites of justice. They must have authority and legitimacy for which they have to command public confidence and respect' (p.51).

It is not only academics who have voiced concern about the state of the buildings which house the civil courts. Members of the judiciary have also expressed anxiety that '. . . the condition of the estate is symptomatic of a culture where civil justice is not driven by the same imperatives as other parts of the system' (Collins 2002: 3). In an Australian context a former Chief Justice has also suggested that there should be a shared vision of courts as being fundamental to our democratic process which requires purpose-built courthouses rather than second rate conversions of office buildings (Black 2009). In the UK it is certainly the case that in addition to their shoddy appearance, a number of County courts are currently held in office blocks rather than buildings with a civic presence.[9] County courts have long been located in the commercial sector of towns by way of recognition of the communities they tend to serve but in the nineteenth century the appointment of County court surveyors meant that these courts were often associated with a particular building type which signalled their civic importance. County courts of the latter part of the nineteenth century benefited in particular from the vision of Charles Sorby who was responsible during his time as County Court Surveyor for upgrading the image of these courts through greater architectural sophistication (Graham 2004). By way of contrast some contemporary County courts such as those at Weston-super-Mare, Ludlow and Altrincham are located above parades of shops. These examples of modern justice facilities provide a stark contrast to the rhetoric of the *Guide*.

In addition to the poor state of the County courts valid concerns can also be directed at the ongoing inclusion of historic buildings in the current stock of justice facilities. Whilst it is easy to support the case for the preservation of many of these buildings as important heritage sites it is not necessarily the case that they remain appropriate locations for the modern trial. Despite the compelling arguments put forward by organisations such as SAVE (2004) and restoration architects (Jones 2009) about the importance of their retention as courts there is a danger that romanticism can cloud sensitivity to outdated symbolism. Serious concerns can legitimately be posed about the extent to which historic buildings, such as those in use in Aylesbury which date from the eighteenth century, sufficiently reflect the principles of accessibility and fairness enshrined in modern design guides. It has been claimed that sensitive restoration of historic courthouses which reflect a mediation between memory, tradition and the contemporary needs of the legal system have taken place (see Jones 2009) but it is not necessarily the case that memorialisation of outdated spatial arrangements with their strong emphasis on hierarchy are appropriate settings for modern justice. A prime example is the Crown Court in Lincoln. This beautiful gothic revival building by Robert Smirke has considerable architectural merit but it remains the case that visitors to the building have to enter a fortified castle complex, pass a former prison in which the segregation system of incarceration was rigorously practised before even entering the courthouse. Heritage clearly has its place but these spatial practices of earlier eras can also serve to bring associations of oppression with them. Alternative readings of these sites remind us that these courts were built during eras in which hanging was common, the ability to defend oneself highly dependent on wealth and women were not allowed on the bench, at the bar table or in the jury box. From an architectural perspective Greene (2006) has also argued that whilst the traditional elements of courthouses such as their massive masonry, blocklike massing, heroically scaled façades rendered in the authoritative language of gothic or neoclassical architecture expressed the power of law, they have done little to communicate concern for the place of citizens in the workings of the process. In his words 'Grandeur has often overwhelmed humanism, and overcrowding and deterioration of the building due to deferred maintenance often amplify the sense of being in the presence of an uncaring, aloof, and unaccountable bureaucracy' (p.65). The common strategy of using outgrown historic courthouses designed for the Assizes for modern Magistrates' courts trying less serious offences than would have been heard at the Assizes.

Reflecting on the courts built since the Beeching Commission SAVE (2004) have concluded that the standard of buildings has generally been dispiriting and disappointing. Architectural successes are seen by some commentators as the exception rather than the norm with most schemes being described as bland and nondescript (SAVE 2004). Particular concerns have been voiced that the introduction of the Private Finance Initiative (PFI) for funding of new courthouses in 1996 has brought down standards of design. The consequent emphasis on a profit-driven ethos has been seen as endangering aesthetic considerations and undermining the notion that public buildings are civic projects which stand for something greater than functional necessity. This is not an issue which the Ministry of Justice has remained insensitive to. The fact that the *Guide* has increased in size considerably since the introduction of PFI can be seen as an attempt to ensure that certain standards in the quality of materials and design are maintained. It is also noticeable that recent design guides are placing much greater emphasis on aesthetic considerations than their predecessors. Moreover the Ministry of Justice has reported that *they are* working with the Commission for Architecture and Environment in pioneering the idea of a design champion for justice in order to ensure that new build meets high design standards (Department for Constitutional Affairs 2004).[10] There is also evidence to suggest that the design and building of the Manchester Civil Justice Centre was viewed as a flagship project for the Ministry and the architects responsible for the design have reported that they were given considerable leeway as regards the design concept which remained relatively unencumbered by questions of cost (Quinlan 2009).

Despite the best efforts of the Ministry of Justice and the partial success of the Manchester Civil Justice Centre it also remains worthy of note that while many of the courthouses of the late eighteenth and nineteenth century were built by leading architects of the day this has not been the case since. Ironically, British architects are designing important courthouses elsewhere. The Richard Rogers Partnership was responsible for the European Court of Human Rights in Strasbourg (1995) as well as the recently built courts in Bordeaux (1993–9) and Antwerp (1988–2006).[11] Norman Foster and Partners have designed the Supreme Court in Singapore (2000–2005) and are now in the process of overseeing the construction of the High Courts of Justice and Supreme Court in Madrid. David Chipperfield Architects completed the City of Justice in Barcelona in 2008 and have since designed the Salerno Palace of Justice in Italy which is due to be completed in 2011.

A closer analysis of the *Court Standards and Design Guide* demonstrates that the goals of accessibility and equality are more frequently discussed in relation to the exterior of courts than they are the interior. Most obvious is the fact that whilst centralised planning appears to have encouraged flatter courtrooms, the practice of segregating and privatising the internal space of the modern courthouse has continued in ways, which I have argued in previous chapters, that remain unacceptable in a modern justice system. The importance of architectural creativity is recognised by the *Guide* but largely limited to the façade and entrance hall. As regards the courtroom, the *Guide* makes it clear that the expectation is that architects will design courts with an eye to tradition and prescribed order. The government have stressed that whilst always looking for innovative designs the scope for innovation does not extend to the freedom to re-engineer core aspects of design (Department for Constitutional Affairs 2005). The assumptions behind such non-negotiable design principles are not discussed further but there is a strong sense of an established order of things which is no longer considered worthy of discussion. This is made particularly apparent in relation to the interiors of courtrooms. The introductory section on the specifications for Crown and Magistrates' courts in the current *Guide* (2010) is worth repeating:

> The courtroom layouts are the result of careful consideration by numerous user groups. They incorporate specific and well-defined relationships between the various participants by means of carefully arranged sight-line, distances

and levels. It has been found that attempts by individual designers to improve on these layouts have rarely been successful and consequently these layouts . . . are to be adopted in all cases. (p.141)

Despite some important shifts in contemporary practice these observations go some way to supporting Graham's (2003) contention that the modern courthouse can be seen as a frozen site of nostalgia in which designers are encouraged to contain any aspirations towards fundamental change in the ways in which people are placed in the spaces of the courthouse. Consideration of these issues raises questions about the ways in which design has been complicit in imposing a particular order on the courtroom which has detrimental implications for contemporary notions of due process. The approach suggested here is a backward looking one in which designers of courtroom interiors are expected to contain aspirations towards progress or change. Viewed in this way courtrooms are seen as having authentic, fixed and unproblematic identities in which the placing of bodies in particular ways is no longer contestable. Is it really the case that the internal design of the courtroom has reached such a peak of perfection?

Writers such as Tschumi (1996) have been sceptical about the opportunities for creative design in such contexts and that for most architects there is a lack of impetus to violate the bureaucratic or political boundaries of their client. In his view historical analysis has generally supported the view that the role of the architect is to project on the ground the dominant images of social institutions rather than to challenge them. Instead they translate the economic or political structure of society into buildings or groups of buildings. Those involved in the compilation of English design guides have acknowledged that the scope for creativity remains limited in justice facilities. Brown (1980a), who chaired the committees which produced the early design guides, has argued that the presence of articulate high status users of the buildings who veer towards a conservative approach to design can make it extremely difficult for the architect to suggest fresh ways of thinking about circulation routes or the public interface with officials. Commenting on current approaches to new build within the Ministry of Justice SAVE (2004) have concluded:

Privately the Court Service admits that it has set itself a very difficult task. Although the political will exists at the highest level . . . the hardest constituency to convince is the senior administrators, trying to balance extremely tight budgets and manage the process as clients. They have yet to be convinced that this new emphasis on design brings benefits that can justify extra financial cost'. (p.17)

The shift towards a risk society would appear to have exacerbated the tendency towards conservative design. Clearly the highly charged and volatile nature of courthouse interactions means that accessibility has to be balanced with issues of protection but Brown (1980a) has also drawn attention to the ways in which a discourse of disorder permeates design negotiation. Reflecting on his experiences of

150   Legal Architecture: Justice, due process and the place of law

consulting with different groups of staff about design he concludes that the police, social services and probation staff prefer to draw strict boundaries around their sphere of operations and that unspoken hostilities are discernible in discussions about the spaces where their respective zones meet. He has concluded as a result that in such a climate the humanist instincts of architectural values can all too easily be dismissed as soft progressive yearnings. Elsewhere he (1987) has argued that the security advisers, with all their paraphernalia of secrecy, obsession and rigidity of attitude have transformed into modern-day 'form makers' (p.33) alongside architects. Heightened security concerns in the aftermath of 9/11 have further encouraged such discourses of containment and surveillance with Smart and Smart (2009) suggesting that security needs have increased tenfold in the last three decades. In some instances, this has led to the placing of the mass of new courthouses further from the street, the use of low walls around the building and the placing of at-risk categories of users deep inside the building (Wright 2004). Architects have commented extensively on such developments and in particular on the impact that airport security measures located in entrance halls of courts have on the public's first impression of the courthouse. Often not anticipated in the original plans it has been argued that:

> . . . the experience of entering the Trial courts is intense, to say the least. Bags and jackets are removed and x-rayed while you are ushered through two magnetometers, wanded and then advised of the very large fines applicable to any unauthorized photography or recording.
>
> (Wright 2004: 78)

Others have drawn attention to the additional danger that bureaucrats tend to favour solutions which make buildings look severe as a way of conveying messages to those inclined to breach security. Reflecting on a court architecture conference in the US, Fulford (1998) attempts to capture a sense of the ways in which tension between designers and security staff continue to manifest themselves in practice:

> Lofty dreams of cultural expression mingled with the cruel realities of crime and punishment [prevailed] at the three-day American Institute of Architects conference . . . The distinction between correct architecture and the architecture of corrections sometimes blurred, though the conference organizers tried to segregate the two themes. At the western end of the Convention . . . 500 or so professionals listened to eloquent speeches about the noble and inspiring face that justice can present to the world through the right sort of architecture. At the eastern end of the same floor they picked up their free copies of *Correctional Building News*, a trade journal advertising precast concrete jail-cell modules, maximum security sprinklers that can't be activated as a lark by people with too much time on their hands, and 'safe, sanitary padded cells'.

Initial images and experiences of the courthouse associated with having one's possessions and body searched are clearly a long way from the notions of accessibility identified as crucial in the *Guide*. Indeed, Loeffler (1999) has argued that if courthouses are to become so fortified that judges and jurors are sequested within their walls then the ultimate cost of such security may begin to outweigh its benefits. Given the lack of people speaking out on behalf of keeping public buildings welcoming it is argued that the time is ripe for a national conversation about the balance between openness and security.

New visions of a democratic future?

Courts are undoubtedly a complex building type in which expectations of progress and stability, power and independence, equality and segregation, security and accessibility must all be played out in the mind of the contemporary architect. These tensions can not always be resolved and are present to different degrees in all courts. Steering thinking towards a more democratic agenda David Tait (2009) has referred to the idea of the contemporary courthouse as a democratic space in a citadel of authority and Dovey and Fitzgerald (2010) have suggested that there is a contemporary expectation that architectural programmes for courthouses must continue to strive towards a range of symbolic and spatial functions which engender respect for justice without mystique, intimidation or exacerbating conflict. Discussions of new approaches to design take on more urgency still when one considers the paradigmatic shifts taking place in the English justice system with its new emphasis on co-operation, rehabilitation and restoration. Initiatives such as the Liverpool Community Justice Centre which focuses on addressing the causes behind a crime provide us with a new model of a court which is socially interventionist rather than remote. In the remaining section of this chapter I turn to look at other innovations in court design which help us to re-think the current canon and find ways of mediating the difficult path between the provision of adequate security to protect those at risk and the need to reinvent the space of the courthouse as an active and dignified public arena.

It follows from the extensive arguments made in earlier chapters that the dock, and in particular the fortified glass dock can be viewed as an anachronism in a modern legal system which aspires to be civilised. While there are undoubtedly a small number of trials in which such security measures are justified it is argued that in the majority of proceedings there is no need to house the defendant in a room within a room. If the United States of America, arguably the most security conscious country in the world, can allow the defendant within the inner bar of the court to sit shoulder to shoulder with their lawyer it seems inexplicable that the same level of dignity can not afforded those who have not yet been proven to be guilty in a UK trial. Reflecting on somewhat unhealthy conservative attitudes of the judiciary in discussion of courtroom layouts Brown (1980a) has suggested that:

> The 'eyeball to eyeball' axial confrontation of magistrate and the accused, with the ability to send the person dealt with directly and immediately 'down',

152   Legal Architecture: Justice, due process and the place of law

> is manifestly a psychological satisfaction geared to deeply held traditional
> beliefs rather than a rational reappraisal of use and need. (pp.1240–1)

Justice Bongiorno's decision to order that a glass dock be dismantled discussed in Chapter four suggests that the judiciary are not always supportive of increasing incarceration of the defendant and that there continue to be opportunities for us to reflect upon such spatial practices.

It follows from the arguments I have pursued in earlier chapters that there is a strong case for members of the public to be afforded greater dignity in proceedings. The practice of containing the public in enclosed galleries with all the audio and visual disadvantages that it brings seem unacceptable in an era in which centralised guidance on court design purports to pay so much attention to accessibility. If the judiciary are to be able to claim that the justice they administer is open it should be treated as imperative that members of the public are welcomed into the court and positioned so that they are able to see the evidence presented and make their own assessment as to whether the outcome reached is fair. Rather than assuming that their purpose is to interfere with proceedings or to threaten other members of the court it should be remembered that it is the attendance of the public which is supposed to legitimate the trial.

A common approach to the problem of ensuring that courthouses are legible to court users in ways which undermine the impression of an impenetrable façade is to use glass in place of walls of brick or concrete. Architects and commissioning bodies have made much of the symbolism of transparency. Glass is seen as an important medium in communicating the openness and accountability of a court to its community and suggests a justice system open to public scrutiny and inclusive of public participation. Frank Greene (2006) has asserted that 'More than simply a glass wall, the phenomenon of transparency, both literal and symbolic, is a defining metaphor of a critical need in gaining public trust in the judicial system' (p.64). Transparency is seen as undermining mystique and inviting public participation by rendering surprises within less likely. The Federal Constitutional Court of Germany in which glass has been used to create actual and metaphorical transparency in the aftermath of the horrors of the Third Reich is just one such example. In this instance Bürklin *et al* (2004) report how the architect carefully avoided the use of awe-inspiring symbols in the courthouse, preferring images of openness which reflected the new rational dignity of a court designed to safeguard civil rights against state intrusion. The Richard Rogers Partnership has also made extensive use of glass in their courthouse in Antwerp which was opened in 2006 shortly after the Belgian legal system had been under attack for corruption following the arrest of the paedophile and murderer Marc Detroux. The building has at its heart a large public space with a roof which is a web of white-painted steel and acres of glass. Commentators have suggested that the building provides a deliberate and stark contrast to the old palais de justice in Brussels built high on a hill to command the city with colonnaded halls of monumental power (Glancey 2006). Glass has also been used to transform the notion of the threshold so that the public sphere of the exterior appears to extend into the courthouse's interior. Greene (2006) describes how the federal courthouse in Boston has a concave glazed public hall that both presents the courtroom entrances to downtown Boston and symbolically brings the downtown into the courthouse by making it visible from within the public hall.

Despite the many benefits of glass it could also be argued that it has negative connotations which are less often discussed. In particular, frequent references to glass symbolising the transparency of justice are in danger of becoming rather hackneyed when public confidence in the legal system is low or reform overdue. In these circumstances glass renders the building dishonest rather than honest. It is also legitimate to ask what is being made transparent. For all the rhetoric of glass facilitating the visibility of law and the legal system most examples cited by architects and policy makers involve the use of glass in public halls and circulation routes used by the general public. But it is the inner workings of the court which remain mysterious and which the public may have a greater interest in seeing. Using glass in public areas can also be viewed as imposing a constant threat of inspection on the public in a society which is becoming increasingly obsessed with their surveillance. These disciplinary possibilities have been recognised by a small minority of architects who have used the material to subvert standard approaches to surveillance in the courthouse. When Bernard Kohn designed the Tribunal de grande instance in Montpellier he ensured that windows from the central atrium into the judicial corridors gave a sense of the enclosed segregation routes behind the walls of the private interior of the courthouse. Even more controversially, at the courthouse in Antwerp, designed by the Richard Rogers Partnership, it is the judicial corridors between the court and the private zone set aside for judges which are constructed almost completely of glass with the result that the judges and public can see each other as they walk along their separate parallel routes on adjoining walkways. The provision of separate corridors for the judiciary reminds us of their special status and the respect which should be accorded them. But the fact that they can be seen by those for whom they act as they go about their everyday tasks, produces a much more powerful symbol of accountability. These examples provide compelling reminders of the fact that architecture can continue to challenge fossilised design templates and find new ways for courthouses to acquire integrity as public spaces.

Elsewhere architects have experimented with the idea of emphasising the public areas of the courthouse in new ways which encourage us to rethink the inevitability of integrating public space with the private zones. At the European Court of Human Rights the Richard Rogers Partnership has separated out the administration wing from the block containing the courtrooms with the result that most of the latter is devoted to public space. A similar scheme has been employed at the European Court of Justice in which two administration towers sit apart from the main court complex. Celebrations of the public arena are also echoed in the design concepts underpinning the South African constitutional court in which the central foyer shown in Figure 7.2 attempts to positively celebrate the right to free assembly in the aftermath of apartheid abuses. These examples remind us that it was once the courts which dominated the courthouse and that a celebration of the areas where the public congregate can re-enter design principles. Increasing use of information technology can also help to return us to a position in which the administration function of the courts can be separated from the place in which all the parties come together. In these ways we can aspire to return to a position in which the private zones of the courtroom do not dominate the 'public' arena of the courthouse.

The Richard Rogers Partnership's 1998 addition to the historic courthouse in Bordeaux has also challenged the extent to which the private zones of an integrated courthouse should remain invisible. In doing so they have been widely accredited as sweeping aside the boundaries of what has hitherto been considered 'appropriate' imagery for a courthouse (Greene 2006). One of the most fascinating aspects of the design is the creation of seven elongated domes clad in cedar which each house a courtroom and sit like a series of organic pods under an undulating copper roof. The idea of the design was to liberate the courtrooms from the 'box' like design of many courthouses and in doing so it has been highly successful in exposing its secrets and rendering the building immediately comprehensible from the outside. The design also separates out the site of the trial from office space and public areas and exposes it freely to the gaze of all passers by.[12]

A similar desire to expose segmentation has inspired Bernard Kohn's designs for the courthouse at Montpellier. Reflecting on the lack of light in many French courthouses and the sense of imprisonment generally engendered by courthouse design the architect was particularly concerned that no one should feel a strong sense of capture or enclosure.[13] His goal was to create an environment in which everyone would feel they could enter, and more importantly exit a courtroom with ease. The result is that the architect positioned the courtrooms as islands within the space of the *salle de pas perdu*. This allows participants who fear the confinement of the courtroom to walk around the outside of the courtroom enclosure and understand the spatial dynamics of their subsequent encasement.[14]

Architects are becoming increasingly sensitive to the psychological aspects of the environment of the courtroom. Innovation in this area is particularly noticeable in Australia where emerging sensitivity to the needs of indigenous peoples has led to significant shifts in practice. It is now widely accepted that aboriginal Australians have suffered, and continue to suffer a high degree of disconnection with European Australians. Lack of attention to the cultural and psychological importance of the land and different concepts of personal space have meant that courtroom architecture has been interpreted as particularly oppressive to peoples who have a strong link to nature. In response to such concerns courthouses such as the Kalgoorlie Courts project in Western Australia are now using outdoor space as a central organising principle in design. Each Magistrates' court in the complex now has its own private courtyard which allows proceedings to take place with direct access to fresh air, light and visual connection to native bush gardens. There is even a courtyard which is large enough to allow proceedings to take place outside in the garden if appropriate (Kirke 2009). Similarly in Port Augusta an outdoor shelter with sweeping views of the local landscape has been created which repeats the roof line of the court complex so that cases can be heard outside (Grant 2009).

These new approaches to the provision of courts encourage us to revisit the ways in which access to open spaces, fresh air and light should be seen as something which all courts users could benefit from and which can be used to dignify the public. Brawn (2009) claims that the importance of spacious public realms, natural light, access to the outdoors and meaningful views out of interior spaces are now informing the best designs especially in the areas adjoining courtrooms where stressed litigants and their supporters are likely to gather. Research and judicial opinion suggests that courtrooms with an external aspect and natural light help to deliver much better trial outcomes in which participants are more relaxed and better able to concentrate for longer period of time (Hockings 2009). In some instances attention to such needs have been recognised in the provision of outdoor areas for participants to have a cigarette before entering the court. In the most ambitious schemes trees and gardens are becoming an integral part of the design of buildings. At Parramatta Justice Precinct a ten-storey glazed atrium with trees exists within the secure area of the court and the architects of the new Supreme Court in Brisbane have also included such a scheme. Similarly, at the South African Constitutional Courts shown in Figure 7.2 the sense of being in a leafy glade is created by a combination of tree-like sloping pillars and partially closed shutters.

For many contemporary architects the challenge of addressing the needs of majesty alongside accessibility are best served by attention to details which humanise the building. Graham Brawn (2009) argues that in the aftermath of the utilitarianism of buildings of the 1970s–1990s the Commonwealth courts construction programme 'reintroduced the power of architectural space to the designer's palette' (p.41). Paul Katsieris (2009) has described how the team of architects at Hassell Architects responsible for the Commonwealth Law Courts incorporated a nine storey-high atrium which could be negotiated in a series of stages by lift or staircase. This has been used to emphasise mass, weight and the gravity of law whilst also including smaller, more manageable elements such as drinking fountains, writing places, telephone bays and gathering places which reduce it to a human scale[15] (see Figure 7.4). On a more modest scale at the Collingwood Community Court the 'modest but daring' open brief has resulted in a long wooden balcony being placed in the area adjacent to the courts which provides nooks and crannies for the defendants, victims and supporters to cluster and claim temporary refuge from the strains of the public arena. Motivated by a similar desire to design in small spaces which could be claimed for temporary refuge Bernard Kohn eschewed the use of straight lines of public seating in his courthouse at Montpellier in favour of kidney-shaped seating

Figure 7.2 Foyer of the South African Constitutional Court © Ben Law-Viljoen, David Krut Publishing.

around which shielded conversations could take place. These stand in stark contrast to the straight and clinical rows of seating installed in the long straight corridors of the highly acclaimed Manchester Civil Justice Centre shown in Figure 7.3.

It seems unlikely that security checks will be abandoned in courts but architects are keen to 'soften' the initial impact of compulsory stop and search procedures and weapons screening technology. Many existing buildings were not designed with such intrusive security measures in mind with the result that the beauty of a public atrium can be upset by the noise and physical presence of security measures. Bottlenecks created by security also serve to militate against immediate enjoyment of a central public space designed to inspire and uplift. Considerable attention is now being paid in the most progressive courthouse designs to the possibility of separating out security from the central public spaces of the building. At the highly successful design for the County Court in Melbourne (2002) and in the Manchester Civil Justice Centre a weapons detection system is placed at the public entrance to the buildings but is not part of the main atrium. Once visitors pass through security they then take a turn into the light-filled central atriums before progressing to the courts inside. At the Commonwealth Law Courts a finger which stretches out at the front of the court allows the security experience to be concluded before entry to the main hall. This allows the

Back to the future: Is there such a thing as a just court?   157

Figure 7.3 Seating in the public waiting area outside the courts at the newly constructed Manchester Civil Justice Centre © Linda Mulcahy.

heavy violence of the law to be metaphorically separated out from its noble aspiration to justice.

If courts are to be seen as public spaces it is critical that the public are consulted about the images of justice they want their buildings to convey. The most recent edition of the *Court Standards and Design Guide* (Her Majesty's Courts Service 2010) suggests that 'users' are regularly consulted but commentators have made clear that the term does not include members of the public as their interests can be conveyed by regular users of the building (Home Office and GLC 1977). Moreover, in the process of designing and consulting about the recently built Manchester Civil Justice Centre the architects were discouraged from even speaking to the public servants working within the building directly and resident judges had minimal involvement in the design process (Quinlan 2009). In other jurisdictions architects have placed considerable emphasis on canvassing the views of members of the public alongside those of staff. Interestingly, widespread consultation with community members about the Collingwood Community Justice Centre led to a design which reflected their preference for a low key entrance hall which had none of the sign or signifiers of a court building and was devoid of security equipment. At another court discussed by Kirke (2009)

Figure 7.4 The Commonwealth Law Courts, Melbourne © John Gollings.

consultation led to separate circulation routes for the supporters of the two parties to the proceedings. These comparative examples encourage a reconsideration of whether the 'well defined' layouts incorporated in the *Guide* can claim to reflect democratic principles of equality and accessibility when opportunities to comment on them is not being offered. Reflecting on his involvement in the courthouses of the late twentieth century Brown (1980b) has concluded in terms which I argue are equally as pertinent today that:

> Whether convincing architecturally, or not, these facades present, in the manner of theatrical tradition, masks which permutate the themes of severity, official-dom, tragedy, neutrality, the benign and the paternalistic. Behind the policy remains effectively the same. In which direction should change be? One possibility is to diversify the use of these judicial pumping stations. All the participants and the general public could join together to re-appraise this building type, seeking changes which might integrate judicial procedures into the community. Civilisation, if it is to be, has no alternative but to be civilized. (p.248)

There are many aspects of the highly praised Manchester Civil Justice Centre which are laudable but one wonders how different the design would have been if members of the public had been consulted. Would the corridors outside each level of courtrooms still have been fitted with metal seating in straight lines which militated against last minute or private conversations with counsel? Would the consultation pods, suspended on high over the multi-storey atrium have had a glass wall exposing those within and machine-like characteristics? Would the quality of the entrance area have been matched by something other than bland fittings in the court?

Conclusion

For most of the history of courts the status, design and architectural merit of courthouses was dependent on local conditions. The wealth of a community, the presence of generous benefactors or levels of competition with neighbouring towns and cities have all determined the state of justice facilities. This way of funding and commissioning legal buildings produced considerable diversity which resulted in majestic, inconsequential, spiritual and degrading palaces for justice. The introduction of centralised responsibility for new build and the emergence of a centralised design guide changed these long-established ways of doing things in the 1970s and policy makers and architects were offered unprecedented opportunities to reflect on what courthouses of the late twentieth century should convey to those who gaze upon them or enter inside. The challenge was boldly embraced by modernists who imagined buildings which reflected new images of law and the post-war social contract. The courthouses which emerged subverted many of the architectural practices which had led design since the end of the eighteenth century. Lines were simpler, courthouses were less troubled by ornament and courtrooms became noticeably flatter. But these seismic shifts in thinking about the design of courthouses are far from being the end of the story of modern courthouse design. In many ways the initiatives of the 1970s are only the beginning of another period in which the challenges and achievements of the classical and gothic building type are subverted, re-imagined and reconfigured in the buildings of today. For architects one of the greatest challenges continues to be how to make the most complicated of building types with its multitude of circulation routes comprehensible to users and how to render a secret building a legible one. Experimentation with rendering the insides of the building visible get to the heart of contemporary debates about public space and the democratic deficit.

Changing concepts of courthouse design have a strong resonance with contemporary debates about the notion of public space. As private shopping centres and entertainment complexes which admit the public vie with the civic spaces of previous generations it is more important than ever before that we see courthouse design as something more than a series of corridors through which the public are herded into their segmented space within the courtroom. Concerns about the emancipation of the spectator which encourage us to focus on the public as more than an inactive onlooker are rife. This is a theme to have emerged in earlier chapters and there are few places in which these issues are as important as in the modern courtroom in which the public and defendant are largely marginalised and rendered passive. Architecture has long been complicit in this process and the conservativism of judges and demands of security consultants has meant that a rethinking of courtroom layouts is rarely on the agenda. One wonders if we were to go back to the drawing board and reconfigure courtroom space in ways which reflect respect for the democratic principles and notions of due process we purport to hold dear whether we would end up with the standard layout of a courtroom prescribed by the *Guide*? Other common law jurisdictions have shown themselves to be much braver in rejecting traditional courtroom layouts in the interests of due process and there is a danger than the inability of the UK to rethink or rediscover what we want the trial to be will make courtroom design no more than a practice in preservation of practices which developed in very different political contexts.

Notes

1 The architectural metaphor of the tree under which African villagers traditionally resolved their legal disputes is particularly important to an appreciation of the conceptual motif of pre-colonial Africa being employed (Le Roux 2004). It has been argued that attempts to move away from patronising and clichéd notions of an 'African' aesthetic have not been wholly successful. The author suggests that despite the buildings' avoidance of triumphal posturing, it perpetuates conventional fictions of nation building by its references to a heroic as well as a repressive past (Freschi 2007).
2 The terms of reference were 'To inquire into the present arrangements for the administration of justice at Assizes and at Quarter sessions outside Greater London, to report what reforms should be made for the more convenient, economic and efficient disposal of the civil and criminal business at present dealt with by those courts and to consider and report on the effect these will have on the High Court, the Central Criminal Court,

the courts of quarter session in Greater London and the County Courts throughout England and Wales'. Described as radical rather than revolutionary the committee was set up in response to the increase in crime combined with generous post Second World War arrangements for legal aid (Chorley 1970).

3 In 1969 the Greater London Council set up a working party with Home Office representation which resulted in a design study for Magistrates' courts and the compilation of a national guide (Home Office and GLC 1977). See further Consultative memorandum on the design of the courts, 1970 and the Crown and County Courthouse Design Manual 1975.

4 Not all jurisdictions followed this lead in developing standardised templates for courthouse design although the issue continues to be debated in jurisdictions such as Australia and France. Despite this, the rise of the centralised planner in the UK and US is such that Wong (2001) is confident enough to make reference to the 'profession' of judicial space management which he goes on to describe as both an art and science. There is now a burgeoning literature on courts house design. See for example Phillips and Griebel 2003; Public Buildings Service 1997a and 1997b; Hurst 1992; Sobel 1993; Wong 2001. For further references see www.ncsonline.org.

5 See further http://www.highcourtchd.gov.in/buildingview1.htm

6 A virtual tour of the High Court can be taken at: http://www.hcourt.gov.au/about_04.html

7 Examples of these flatter courtrooms can be seen in Pinfold 1963; Harris 1959; Barker 1964.

8 The need for natural ventilation has clearly been taken very seriously by successive government departments responsible for courthouse design since the 1970s. Henocq (1999) reports that the PCA commissioned Cambridge University scientists to experiment with various designs for air flow.

9 Examples of such County courts include those at Gateshead, Wellingborough, Wrexham, Lincoln, Hertford, Thanet, Basingstoke, Barnstaple, Aldershot and Farnham and Stockport. Pictures of these courts can be seen at www.hmcourts-service.gov.uk.

10 In the interests of encouraging competition honorariums are now being paid to architectural practices who finish second and third in PFI bids (SAVE 2004).

11 The Richard Rogers Partnership did submit an entry for the Manchester Civil Justice Centre competition and were placed on the shortlist of three but in the event Denton Corker Marshall were given the commission (Denton 2009).

12 Images of the Bordeaux courthouse can be seen at http://www.rsh-p.com/rshp_home.

13 Private conversation with the author during the Courthouse of the Future tour 2008.

14 This information was gleaned during a visit to the court by the author in 1998 when a group of delegates from the Court of the Future Network was shown round the courthouse by the architect.

15 A virtual tour of the Commonwealth Law Courts can be taken at http://www.abc.net.au/arts/architecture/interact/tours/court.htm

Chapter 8

The dematerialisation of the courthouse[1]

Introduction

This chapter draws attention to the ways in which the increasing use of technology in the courtroom has the potential to disrupt traditional concepts in court design and considers the claim that the courthouse, as the prime site of adversarial practice is in danger of being dematerialised. One of the key issues to be explored is whether or not this reflects progress in our thinking about the interface between buildings and the trial. I argue that the promises of technology are contradictory and confusing. The new devices available to courts have the potential to disrupt some of the practices discussed in this book so far and to liberate participants in the trial from the strictures of tradition. At the same time, they can serve to radically alter the place of architecture, ritual and ceremony in preparing participants for their role in the legal arena. These changes call for more extensive public debate than has so far taken place. A key question here is whether the sort of trials now being imagined in which architecture appears to play an increasingly minor role render the process an 'inauthentic' legal ritual? It is certainly the case that in the new digitally mediated and virtual environments being envisaged, and slowly realised by policy makers and 'techno-evangelicals', the character and content of legal discourse is being transformed. Encounters within the courtroom are in danger of becoming sanitised as participation in the trial becomes akin to a fleeting televisual encounter. Conceptions of the court as centrally located, locally anchored, spatially discrete and architecturally symbolic are also set to change. The result is a radical reorganisation of the challenges that 'spatiality' poses to law.

It would seem that we are to be faced with a number of challenges to received concepts of the trial, all of which are worthy of extensive debate. The physical boundaries of the court have been breached in a number of new ways in recent years which render the surveillance technique of the judicial sightline rather meaningless. Real time transcription funded by one or both parties to the trial has made some participants in the trial invisible to the judge. By allowing a script of the proceedings to be transmitted to back room lawyers located miles away, invisible participants in the prosecution or defence case can identify inconsistencies which can be communicated to counsel via email and seized upon in the course of ongoing cross examination. 'Live link' allows the evidence of witnesses to be transmitted from physical locations outside of the court without them having to enter the courthouse or courtroom.[2] Filming of courtroom proceedings also allows images of the court to be consumed without the necessity of attending the public gallery. In their different ways these practices expand the concept of the witness stand, the area set aside for lawyers and the public gallery to a secure witness suite which may or may not be in the physical vicinity of the court, the office of a well-resourced law firm and the homes of the public. As the walls of the courthouse are violated in the myriad of ways made possible by technology important questions about the unity of the drama of the trial and who produces images and controls the information highways which stretch the court beyond brick and mortar need to be addressed and their implications interrogated.

The purpose of this chapter is not to bemoan the impact that technology is having on contemporary society. It is accepted that technology can enhance democratic participation by promoting discussion amongst the marginalised and geographically remote. Nor is an idealised vision of the trial as a just and fair method for the resolution of disputes entertained. As I have suggested in earlier chapters participation in a trial can be degrading and inhibiting, cross-examination brutal and the aftermath of proceedings traumatic. What is argued is that beneath the inadequacies of adversarial adjudication there may continue to be some design ideals enshrined in court practices which we want to retain in a technological age. In an era in which reform of the litigation system has been fuelled by considerations of efficiency and proportionality, it is argued that we are in danger of forgetting that requiring the physical presence of people in a special building continues to have considerable cultural resonance.

In the sections which follow these issues are discussed in relation to one particular development, the giving of evidence by live video link (live link). The argument is presented in two parts. In the first section of the chapter, I consider the ways in which technology disrupts the experience of the trial when participants are able to 'attend' court by means of live link. The second section reflects on whether the trial loses its potency as a public ritual when stripped of its physical surroundings. In earlier chapters I have rehearsed the many ways in which the positioning of walls, windows and stairwells have been used for many hundreds of years to prepare people for the drama of the trial and to designate place and hierarchy at law's altar. The aim here is to unravel the implications of modern processes of fragmentation and loss of face-to-face transaction. The increasing use of live link forces us to reconsider the spatial organisation of social relations in the justice system and to rethink the unity of space and place assumed by our forebears. A major concern of this chapter is whether, as the old order of exclusive locations dissolves, inadequate discussion of the implications leaves justice placeless and disorientated.

164   Legal architecture: Justice, due process and the place of law

Technology in the court

The quest for greater use of new technologies in the courtroom has recently grown in momentum and a number of bureaucratic structures have been set up by government departments to ensure that use of electronic media is more widespread.[3] Progress in the introduction of technological aids in the UK litigation system has undoubtedly been very slow when compared to other jurisdictions, most notably Australia, Singapore and the US.[4] Regardless of this, it is the vision of the future now being imagined that provides the focus of discussion here. Ministers have suggested that the government has been preparing for a twenty-first century in which the courts in England and Wales will be amongst the most technologically advanced in the world (Department for Constitutional Affairs 2000). Reflecting on such developments Davidson (2000) has asserted:

> The vigorous technological shake up which is about to hit the profession is comparable to the meteorite millions of years ago which devastated the dinosaurs ... The sophisticated technology available today requires the legal profession to examine the basic tenets of justice. It is a jurisprudence professor's dream. (p.6)

To date, technology has been extensively used to increase the efficiency of routine tasks in the litigation process. Recent innovations in litigation-support technologies include electronic storage of files, electronic filing of proceedings, the availability of electronic copies of standard form documents, full text retrieval, increased use of document image processing and links between e-diaries and automatic listing of the trial. The expectation is that in time judges will be able to gain access to electronic versions of all the important documents relating to a case and communicate with a range of professionals prior to the trial from a remote location and within a safe network.

Other developments are less innocuous. Rather than altering *how* things are done by increasing speed or reducing cost, these new practices alter *what* is done in the trial. Real-time transcription now alters the pace and experience of the trial as advocates no longer pause for judicial note-taking. The orality and central role of performance in the trial is further compromised as judges are presented with a written transcript of testimony seconds after it is delivered in person before them. Commentators have drawn particular attention to the increasing use by resource-rich lawyers of 'smart screens' or interactive whiteboards to enhance the presentation of an argument (West-Knights 2005).[5]

Several pieces of legislation now allow evidence to be given by live link and the introduction of such provisions has been fuelled by two arguments which need to be clearly distinguished. The first of these relates to the use of live links to shield witnesses considered to be deserving of protection by the State in criminal trials. In exercising their new-found discretion as to whether to allow live link evidence to be given, the courts have generally been required to look at whether its use is in the interests of the efficient or effective administration of justice. The Youth Justice and Criminal Evidence Act 1999 has shifted the emphasis of statutory approaches and now requires that, save in exceptional circumstances, the evidence of witnesses under 17 years of age in relation to sexual or violent offences *must* be given by live television link or video-recorded interview.[6] The change in focus is such that it has been claimed that the legislation heralds a new era and has introduced reforms of the orthodox adversarial trial which constitute the most radical in the common law world (Cooper 2005; Hoyano 2007).

It is essential to the arguments being pursued here to stress that the use of live link in such circumstances is best classified as *exceptional*.[7] The scenarios imagined involve cases in which abrogation of the defendant's expectation that those who speak against them will appear in the court is 'trumped' by witnesses who are children, vulnerable adults, feel intimidated or are in fear or distress.[8] It is argued that the broader interests of justice are served because, if they were unable to use live link, many witnesses in this position would refuse to give evidence at all (McEwan 2000).[9] For this reason live link initiatives have proved popular. Research suggests that 33 per cent of witnesses using special measures said they would not have been willing to give evidence without live link (Hamlyn *et al* 2004; Kitchen and Elliot 2001) and that the police, Crown Prosecution Service and court staff view live link as either effective or very effective (Burton *et al* 2007; Burton *et al* 2006a). In their empirical study of measures for witnesses in need of protection, one research team found that the highest number of applications for special measures were for a live link and that the popularity of this medium was likely to continue unabated (Roberts *et al* 2005; Burton *et al* 2006c).

It has also been argued that when vulnerable witnesses do give evidence through live link it is likely to be of a higher quality because they are less frightened and inhibited than if they were in the physical presence of the accused. Research on live link conducted in a Scottish context found that child witnesses who used this medium were less likely than those who appeared in court in person to cry, report fear or describe the trial as unfair (Scottish Government 1995). Significantly, the use of video links in such cases actually *reinforces* the value of traditional practices by trying to get as close as possible to live evidence in circumstances in which it would otherwise be impossible. In this context, live link has been widely hailed as superior to other attempts to protect vulnerable witnesses, such as the use of screens in court, the submission of written testimony, pre-recorded evidence and the use of intermediaries.[10] In part, this is because it allows for more unity of time, place and action than would otherwise be possible and provides opportunities to judge demeanour. Davies' (1999) evaluation of the impact of television links on the reception of testimony suggests that, while jurors did demonstrate a preference for 'live' testimony, the use of live link did not influence their decision-making and appears to have had no overall impact on the proportion of guilty verdicts delivered (see also Taylor and Joudo 2005). In view of these arguments it would seem that the need for live link in cases involving vulnerable, intimidated or incapacitated witnesses is compelling.

166   Legal architecture: Justice, due process and the place of law

The second context in which the use of live link has been justified has no connection to the particular problems experienced by vulnerable witnesses and raises more fundamental questions about our expectations of other types of witness. The practice in question is the use of live link in circumstances in which it can only be justified in the interests of efficiency. This category of use replaces traditional testimony even in circumstances in which there is nothing in the nature of the proceedings that would prohibit a witness from giving evidence in person. Whilst the expectation is that live links should only be used where 'appropriate',[11] arguments about cost savings have clearly proved compelling to policy makers. Support for extensive use of live link has, for instance, been made in relation to perfunctory hearings. In *Black* v *Pastouna*, Lord Justice Brooke argued that applications to the Court of Appeal which are likely to last half an hour or less do not warrant an oral hearing and that in such circumstances the court expects the parties or their advisers to use live link facilities if it would save costs.[12] The Police and Justice Act 2006 also allows live link to be used to replace 'routine' proceedings. Ministers have estimated that if defendants consent to appearing in court via live link it could reduce the 235,000 journeys which remand prisoners make to court each year (Hoon 1999). The 2006 Act also allows for live links to be used in certain sentencing decisions. In pursuit of such goals, video links between criminal courts and prisons have been installed in 30 Crown court centres and 170 magistrates' court centres (Brooke 2004). In the most radical move away from the spatially anchored trial to date the Justice Secretary announced the introduction of the first virtual court pilot scheme for dealing with minor offences with minimal delay in May 2009. The Virtual Courts scheme involving 14 courts involves defendants having their cases heard at a magistrates' court, via secure video link from a police station, within hours of being charged and if a defendant pleads guilty, could see sentencing handed down on the same day. Jack Straw has claimed that the scheme could save the Ministry of Justice £2.2 million in the first year of operation and will lead to increased satisfaction in the public's experience of the trial.[13]

The pace of change has been even faster in the civil sphere. Since video links were used for the first time in an English civil court in 1992,[14] it has become a widely accepted practice to make use of this technology (Archer 1992; Sage 1995). Indeed, commentators have suggested that the launch of the Government consultation paper *Civil Justice: Resolving and Avoiding Disputes in the Information Age* heralds an era in which live link will become the norm. When the libel case *Polanski* v *Conde Nast Publications Ltd* was heard in the Court of Appeal in 2003, Jonathan Parker LJ could find no difficulty with such a trend. In his words:

> The improvements in technology are such that, in my recent experience as a trial judge, the giving of evidence by video conference link (VCF) has become by 2003 a readily acceptable alternative to giving evidence in person, provided there is a sufficient reason for departing from the normal rule that

The dematerialisation of the courthouse   167

witnesses give evidence in person before the court. In the ordinary run of a case, a sufficient reason may easily be shown. If there is sufficient reason, then even in cases where the allegations are grave and the consequences to the parties serious, the giving of evidence by VCF is now an entirely satisfactory means of giving evidence in such cases . . .[15]

Although the decision of the Court of Appeal was overturned when heard in the House of Lords, this was not because of any unease about the use of live links. Indeed, the Lords readily accepted the trial judge's assertion that evidence given by live links could take place as naturally and freely as when a witness is present in the courtroom and did not cause any prejudice.[16] Baroness Hale went as far as to say that 'new technology such as VCF is not a revolutionary departure from the norm to be kept strictly in check but simply another tool for securing effective access to justice for everyone' (para 80).

Significantly, the notion of the defendant's 'right to confrontation' has proved to be of limited value to those interested in challenging the increasing use of live link.[17] This 'right' has long been recognised in the common law and has a doctrinal foothold in Article 6 of the European Convention on Human Rights which expressly safeguards the right to examine witnesses. However, it has become far from synonymous with face-to-face confrontation in open court. In particular confrontation has been limited in a number of significant ways by Strasbourg jurisprudence which Jackson (2005) argues increasingly favours a participatory form of legal proceedings which draws on precepts from common *and* civil law jurisdictions. While considerable support continues to be given to the notion that convictions should only be based on evidence which has been offered for challenge the European Court of Human Rights has been reluctant to recognise that such challenges need to occur in a public trial, in a particular place or in person. So, for instance, cross-examination does not have to be face to face when live link is available. In this context the competing rights of witnesses in need of protection have been taken very seriously and balanced against those of the accused which have not been treated as absolute. The common law expectation that confrontation should occur during public proceedings has also been undermined as pre-trial evidence heard by investigating magistrates and subjected to defence questioning in private has also been recognised as admissible at trial (Jackson 2005).[18] It has even been determined that evidence of those who have never been confronted may be taken into account as long as it is not decisive. Taking into account a host of other exceptions such as those relating to hearsay, McEwan has surmised that it is something of a 'lukewarm' doctrine in an English context which generally generates more heat than light.[19]

Whilst it has been suggested that the convenience of live links should not dictate their use,[20] those at the forefront of technological developments have an even bolder vision of the possibilities of live link than suggested by the courts in *Polanski*. Pawley (1998) predicts the end of architecture as anything other than a heritage or tourist industry and has argued that the importance of buildings today

168 Legal architecture: Justice, due process and the place of law

is not as monuments but as terminals for information. Such ambitions are best illustrated by the Courtroom 21 Project, a joint initiative of the Law School of the College of William and Mary and the National Center for State Courts in the US.[21] The project team has developed two portable courtrooms equipped with every known technological aid to litigation in order to demonstrate the possibilities which electronic resources offer.[22] Courtroom 21 has already been used in one virtual trial in which two judges appeared 'remotely' (West-Knights 2005). Lederer (1991), who is closely involved with the project and has written extensively about it, has gone so far as to suggest that the twenty-first century will be the age of remote appearances:

> As high technology courtrooms become 'traditional' the legal system will have to cope with an increasing number of remote appearances by trial participants. The true virtual courtroom with its elimination of distance and expense will tempt us. Bar, bench and society will have to decide what a 'trial' means and what 'due process' connotes in a technological age.

In a UK context, Roberts *et al* (2005) have echoed this vision and encouraged us to take the 'special' out of special measures in the criminal justice sphere:

> If special measures *are* an unmitigated 'good thing' with no undesirable side effects, why stop at especially vulnerable or intimidated witnesses as meritorious candidates for assistance? After all, few people positively relish the prospect of searching forensic examination in the witness box, and those who have sampled the experience self-report alarmingly high levels of anxiety, fear and feelings of intimidation.

In a similar vein, Widdison (1997) predicts that even the receipt of disputed testimony via live link will become the norm within the next 10 years. It is certainly the case that predictions about the number and characteristics of courthouses in the future now tend to assume that courtrooms will be smaller and high tech with fewer cells beneath them (Her Majesty's Court Service 2009). Widdison (1997) encourages us to confront the challenges and dilemmas of live link by considering what, other than tradition and cultural conditioning, stands in the way of extending virtual pre-trial proceedings to first instance hearings, re-heard cases and appeals. It remains to be seen whether we are witnessing the beginning of the end of the tradition in which the parties in a trial are physically present in the same place or only the latest phase in procedural adaptation and evolution (Roberts *et al* 2005).

Debate about the increasing use of live link raises a number of issues about the privileging of certain types of witnesses, the rights of the defendant and the significance of demeanour. All of these are undoubtedly important but my focus here is on one particular aspect of these changes that has received very little attention to date; the implications of these developments for court architecture and designs and the significance of the vanishing courthouse on the dynamics of the trial.

Law's aesthetic

As I have demonstrated in earlier chapters considerable thought has been given to the physical surroundings in which evidence has been given for centuries. Resources have been lavished on courthouses and courtrooms in order to stress that the administration of justice plays a central role in the civic landscape and imagination. Choices of site and style commonly reflect assumptions that the meting out of justice is a special function of the State and not something which can be dispensed just anywhere or on a whim. In his discussion of the importance of place in the adjudication of commercial disputes in medieval Italian cities, Burroughs (2000) reminds us that:

> We readily think of legal transactions as ritualised events performed according to a social script conferring authority and resulting in documents that guaranteed assent through familiar means, notably official language and format and the signatures or marks of protagonists and witnesses. The force of a legal transaction, however, might also derive from the setting of its enactment, which might confer, for example, prestige or charisma, or offer a symbolic testimonial function to a wider audience than the witnesses registered in the documents. (p.65)

Since custom-built courthouses became common from the eighteenth century onwards the places where evidence is delivered have become much more clearly associated with a particular aesthetic that serves to reinforce the idea that they are unusual places. I have argued in Chapter six that this concept was particularly well appreciated by Victorian architects who were responsible for the building of elaborate public monuments to law, such as the Manchester Assize court, Victoria Law Courts in Birmingham, St. George's Hall in Liverpool and Leeds Town Hall discussed in Chapter three. Close attention to the surroundings in which evidence is given continues to this day and in recent decades has become much more prescriptive. This is clear from the detailed guidance on the physical appearance and layout of Crown, County and Magistrates' courts contained in the *Court Standards and Design Guide* (Her Majesty's Courts Service 2010). This 1016-page document contains a series of complex illustrations and specifications which prescribe in minute detail how the internal space of all courthouses should be configured. Its intricate diagrams and maps are presented as a definitive plan of how space in the courtroom is to be allocated. Even the most cursory reading of the text reveals the *Guide*'s concern with the minutiae of courtroom signifiers. So, for instance, it contains detailed guidance about the materials which should be used to build and fill the court as well as their quality, size and position. It has sections on signs, safety, air, water, acoustics, furniture and furnishings, finishes and materials, alarms, information technology, and sustainable development. We are informed of the correct height, width and depth of the witness box, its position in relation to the judicial dais, jury, advocates and defendant and the preferred materials to be used in its construction. In short, the distribution of bodies, walls, lights and gazes into particular arrangements reveals conceptualisations of the relationship between the witness and other actors in the trial which have been thoughtfully determined over hundreds of years.

While conceptions of the role of law in society may change over time, the central argument that buildings are viewed as an appropriate medium through which to represent prevailing ideals of justice remains intact. In a UK context, the *Guide* makes clear that the baroque designs of the eighteenth to early nineteenth century which placed emphasis on drama, awe and the instillation of emotion in court users have been replaced by a preference for court buildings which are accessible. But significantly the essential emphasis on the promotion of an aesthetic remains:

> Court buildings need to be seen to be there and seen to be public, authoritative and important in society, whether an individual has reason to use them or not. Unlike our other buildings, therefore, court buildings cannot be hidden away on an industrial estate or be anonymous suburban office buildings, nor can they be seen to be high security compounds or visually oppressive. (p.19)

Though rarely written about and often assumed, it seems clear that the physical surroundings in which evidence is given plays a critical role in reinforcing the importance of the trial and the role of state-sanctioned adjudication in our society. As a result it seems legitimate to ask how profoundly the loss of such signs and signifiers alters the giving of evidence when it is delivered by live link from a remote location. It would seem that there is still considerable support for the notion that the administration of justice and the respect in which the justice system is held are best promoted when trials are located in distinctive, prominent and dignified buildings.[23]

Challenges to these assumptions

The Civil Procedure Rules (CPR) governing live link do not completely ignore the importance of setting in preparing participants for giving evidence. They make clear that live link evidence should be as close as possible to evidence given in open court and it is recognised that there is a need to contribute to a sense of gravity in proceedings. So, for instance, when it is necessary to remove the judge from the courtroom to a video suite in order to hear live-link evidence arrangements are expected to be made, if practicable, for the royal coat of arms to be placed above the judge's seat.[24] Following conventions on 'sightlines' in courtrooms, it is also required that cameras be installed in such a way that the judge is able to see everyone in the room; and the witness any person making statements or asking questions. The special respect afforded the judge in the physical court is also reflected in the expectation that he or she enters the local site for video transmission after all the other parties and leaves before the on-line link is broken.[25]

The dematerialisation of the courthouse 171

Figure 8.1 Video suite in Manchester Civil Justice Centre © Linda Mulcahy.

However, other than these general guidelines there appears to be little expectation that the environment in which evidence is given by live link in these new legal domains be treated as special in any way. The result is that many of the video suites provided within courthouses such as those in the much-hailed Manchester Civil Justice Centre shown in Figure 8.1 are bland and perfunctory. The same can be said for the video link suite shown in photographs and video clips announcing Jack Straw's new Virtual Court initiative.[26] In her work on the 'Gateways to Justice' project funded by the Australian Research Council in which a host of video suites within Australia and other jurisdictions have been visited and evaluated Emma Rowden (2009) has argued that:

> One could argue that remote participation spaces are really an extension of the courtroom – and yet, more often than not, the places in which remote participants find themselves located are small, windowless, no larger than a broom closet, bland, with only a chair and the video-technology itself – and the UK Virtual Court Pilot it would seem, is no exception. In one of the few studies to ask participants what they thought of remote witness facilities, repeated phrases include 'it was like a cupboard', and 'it was like a box'.

One Australian Judicial Officer cited in her study reported that being in a remote court space is currently an experience more akin to being in 'a transit lounge' in which participants were bussed in and bussed out and barely remember what they said or did. These arguments seem credible when one considers that proponents of the Virtual Court pilot scheme in the UK have rather proudly cited the case of a defendant charged with being drunk and disorderly appearing at court via video link two and a half hours after being found in a state of inebriation. We are told that the defendant in question pleaded guilty and was sentenced less than three hours after being charged.[27]

In other cases in which external accommodation for live link broadcasts is prescribed the accommodation and its location is far from neutral and may even be designed to evoke fear. The expectation that the transmission will sometimes take place when the accused is in detention in a police station[28] has since become an expectation in the Virtual Court pilot. This move has all sorts of implications for debate about the efficacy of the dock raised in Chapter four as under this scheme the remote witness facility within a police station is transformed into the dock.

Other scenarios imagined by legislators allow for evidence to be given from any convenient location, regardless of whether it has been built as a justice facility. So, for instance, in *Polanski* v *Conde Nast Publications Ltd* (2003) the claimant asked to give evidence from his hotel room.[29] Moreover, guidance on the Police and Justice Act 2006 suggests that a witness might be able to give evidence from their place of work rather than have to travel to a courthouse. Significantly, such protocols are silent on the issue of how, and indeed whether, the facilities used for live link should be rendered suitable for the function to be performed within them with the consequence that the court service loses its control of setting and its ability to set a tone. These provisions appear to pay little heed to the goal of rendering live link as close as possible to live performance within the courtroom. The transformation is also much more radical than a shift from an imposing to a neutral location. Instead, the importance of architecture and design is marginalised, if not completely denied. It would seem then that live link has the potential to place key actors in the trial outside of the courthouses that have become synonymous with State-sanctioned adjudication and replaced them with the mundane. Commenting on her impression of two of the most technologically advanced courts in the world, the International Criminal Tribunal for the Former Yugoslavia and Vancouver Courtroom 20, Radul (2007) suggests that the prevalence of screens disrupts traditional symbols of authority and transforms the courtroom into a space which is akin to 'the board room of a not very creative (and perhaps not very successful) corporation'.

The lack of accord with performance in the courtroom is further demonstrated by the fact that when evidence is transmitted from a witness's home, hotel room or workplace, their movements are kept to a minimum, their surroundings remain constant. Their 'arrival' in the courtroom is activated by the transfer of their image along an information highway but they remain physically static. Preparation for

The dematerialisation of the courthouse 173

the giving of evidence remains a cerebral one and the experience of the court is minimalised. Live-link witnesses remain unaffected by the influence of court architecture or ornament. The ritual of a journey to the court and away from it are denied them. In her call for research into the impact of video links on participants of the trial the architect Rowden (2009) supports this contention and argues:

> . . . with the *structural* change of removing the defendant from journeying through the more formal and *symbolically*-laden setting of the courthouse, the participant might not pick up on the seriousness of the occasion, behave inappropriately, which may in turn impact adversely on the outcome of their case. Even the most humbly furnished Magistrates' courtrooms and court buildings can have, by way of their structural features and spatial syntax, important environmental cues that may positively prepare and influence a more appropriate demeanour and behaviour.

It would seem that when the court enters their space at the will of a technician and just as easily vacates it law comes, it goes, but it is constantly elsewhere.

So what?

It was suggested in the introduction to this chapter that the giving of evidence by live link seriously disrupts traditional notions of the trial, but it could be argued that there is general merit in disrupting the traditional space–place dynamic in adversarial systems. Perhaps the mundane is preferable to the fear-inducing courtroom and more likely to reflect contemporary aspirations to 'accessible' justice. As I have argued in previous chapters the way space has been used in the courthouse has been fundamental to the exercise of power by the privileged. Studies have repeatedly drawn attention to the difficulties witnesses have in appearing confident while describing intimate experiences to others across a large intimidating courtroom.[30] Carlen (1976) reminds us in her study of magistrates' justice that, as well as being designed to uplift, architecture can be used to inhibit understanding and discourage participation, while also engendering fear and respect. In short, the exploitation of courtroom space can have a paralysing effect on those who are not regular users of the court system and can serve to contribute to a ritualised stripping of dignity. The journey from courthouse threshold to courtroom can be conceived of as one which engenders dignity; it can also be seen as one in which the coercive power of the State is signalled by security checks and the lack of a safe haven. This is as true of well-maintained and elegant courts with their grand entrance halls and ornate fittings as it is of the shabby stairwells of the type which lead to the public galleries in courts in the Old Bailey and Aylesbury Crown court. Seen in this way, the space in a courtroom becomes a particular articulation of social, cultural and legal relations in which some actors are privileged and others disempowered. When seen from this perspective giving evidence from a familiar space, such as the home or workplace, seems attractive. Even hotel rooms

174   Legal architecture: Justice, due process and the place of law

and perfunctory video suites have the attraction of being relatively anonymous and neutral. It could be that the increasing use of live link offers opportunities for witnesses to avoid the imperatives of law's temple and to liberate themselves from the constraints that such materiality and symbolism impose on us.

But debate on the issue might also prompt us to ask new questions about the buildings which provide homes for trials. Discussion of the impact of new technologies in the courtroom is likely to, and should, force us to address fundamental issues as to whether prevailing configurations in the courtroom and its environs continue to be vital in the modern legal system. Unless we encourage this debate, live link could, somewhat ironically, be seen by default as a further step in encouraging the segregation and separation of participants in the trial into neat and easy to manage constituencies. However, it is also important to recognise that there are rarely explored dangers in the outright rejection of traditional uses of space. In the final section of this chapter I turn to consider the impact of the arrival of live link on the cultural significance attached to physical presence at the trial[31] and the 'authenticity' of the trial as a social ritual. Three particular issues are of relevance in this context. What impact does the removal of the material trappings of the courtroom and courthouse have on the experience of preparing for and giving evidence? How is the concept of the trial altered by circumvention of the obligation to be in the presence of those you accuse or implicate? Does physical presence remain important whatever the characteristics of the building used to house the trial.

Meaningful encounters?

Much of the discussion about live links to date focuses on the ways in which it impacts on the rights and experiences of key actors within the trial, such as witnesses, lawyers and defendants. It is noticeable that there is much less focus on how it impacts on the experience of the trial as a group experience or public ritual. Yet these issues are far from unimportant. It is argued here that if trials are to be seen as legitimate public events they have to continue to have meaning as human encounters. There is a danger that, rather like Lady Hale's assertion in *Polanski*,[32] techno-evangelicals are too quick to deny fundamental differences which inevitably occur in the transformation from architectural code to video code. Yet it seems that these need to be unravelled if we are to be clear about what the trial is becoming. When one compares the minutiae of the guidance on courthouse design provided by the *Guide*, with the lack of reflection about the impact of these technological changes on the unity of time, place and action in the trial contained in live link protocols, the limitations of current debate become all too clear.

Does there continue to be something special about physical proximity which should encourage us to maintain face-to-face contact for non-exceptional witnesses whatever alternatives technology offers us? Is improving efficiency a sufficient justification to disrupt the dynamics of one of society's most important

The dematerialisation of the courthouse 175

public events? The requirement of face-to-face interaction has a respectable heritage. There has long been a preference for orality over written evidence within common law jurisdictions, despite there being an historical precedent for the latter in the English Courts of Equity. Moreover, it could be argued that the right to confrontation and jurisprudence of Article 6 hint at much more fundamental ideas than protecting a defendant's right to cross-examination of witnesses or the judging of demeanour. Blumenthal (1993) alerts us to a need which is just as important when he draws our attention to the fact that it is recorded in Acts 25.16 that the Roman Governor Festus declared that it was not the manner of the Romans to deliver any person to die before seeing the accused face-to-face and asking him to account for himself (Friedman 1998). At certain times in our history the attendance of spectators at a trial has actually been conceptualised as a civic duty and that this expectation was used to confer legitimacy. It would seem that the physical presence of the parties and spectators in the courtroom has long assumed a special importance in our culture. Is this grounded in anything other than pragmatism or historical accident? Has the trial become synonymous with a gathering of people only because of a lack of alternatives?

Generations of lawyers writing in support of face-to-face oral hearings in cases other than those involving vulnerable witnesses have argued that physical attendance has immense practical value which makes it easier to evaluate the nuances of body language and demeanour. Bello (1999), for instance, quotes one lawyer as saying that the virtual trial is unlikely to prove popular with lawyers because of the need to see the sweat on a person's brow when you cross-examine them. By way of contrast, live link witnesses appear to spectators in the court from the other side of a flat glass screen. Witnesses using the medium are instructed to avoid unnecessary movement for fear that it will have an adverse effect on the quality of transmission. Even Lord Justice Brooke, who has been a keen advocate of technology in the courtroom, has argued that 'we still have a good deal to learn before we can be sure that the lawyer–client interface is not being dangerously impaired in the quest for economy and efficiency'. He goes as far as to say that many Crown court judges believe that the jury has acquitted in cases where they might well have convicted because they were unable to establish the same rapport using live link. Others have suggested that the sense of emotional distance promoted by live link makes it easier to lie during cross-examination (Costigan and Thomas 2000). Important though these claims are, it could be argued that the preference for physical presence of most witnesses is, and should be, based on more than the ability to judge demeanour. In part this is because sufficiently serious reservations have been articulated about the extent to which the demeanour of a nervous or frightened witness can ever be a true reflection of their state of mind and credibility.[33]

More significant in the present context is the importance of physical presence *per se*. It could be argued that the expectation that a person makes their accusation in the presence of the accused in a setting designed for public gatherings speaks to a society that has an active public sphere and a sense of the collective. By way

Figure 8.2 McGlothlin Courtroom at the Center for Legal and Court Technology. © McGlothlin Courtroom (Center for Legal and Court Technology).

of contrast when evidence is transmitted from the home, workplace or hotel room it suggests that the performance of civic duty in public has become an inconvenience. If, as Sennett (1974) has so eloquently argued, the impact of late industrial capitalism has been to erode the sense of public life as a morally legitimate sphere, those of us who care to retain a strong sense of meaningful accountability should remain sensitive to attempts to undermine the value of physical proximity. Such sentiments have also been neatly encapsulated by Mitchell (1996):

> What does it matter? Why should we care about this new kind of architectural and design issue? It matters because the emerging civic structures and special arrangements of the digital era will profoundly affect . . . the character and content of public discourse, the forms of cultural activity, the enaction of power, and the experiences that give shape and texture to our daily routines. Massive and unstoppable changes are underway, but we are not passive subjects powerless to shape our fates. If we understand what is happening, and if we can conceive and explore alternative futures, we can find opportunities to intervene, sometimes to resist, to organise . . . and to design. (p.5)

One experience promoted by face-to-face encounters in the courtroom is a sense of drama. The idea of the trial as theatre has become somewhat hackneyed and it

The dematerialisation of the courthouse 177

is often used to promote the idea that the adversarial trial and the role of lawyers place the investigation of 'the truth' in peril. This view of the notion of drama is one which assumes that it has a distorting influence on perceptions rather than one which sees performance as a way in which significance can be attributed to acts and words. But understanding the trial as a dramatic performance also has the potential to help us unravel why the face-to-face dynamic is unique. It is possible that a court which lacks a properly theatrical aspect is not a court at all. If we accept that the experience of watching a film in a cinema is a radically different experience from live theatre, why are we not more alert to the difference between live performance and images conveyed on a flat screen in the courtroom?

Looking at the issue from the perspective of a performance artist, Radul (2007) offers an important insight into what the essential difference might be. Firstly, she draws attention to the fact that the screen on which video link evidence is played in the courtroom places a barrier of glass between the witness and other participants that does not exist in the traditional trial. In her view the glass excludes the object from the consumer and produces a dialectic of inclusion–exclusion, presence–absence and divider–connective. Video link presents a framed view of things happening beyond the glass which focuses on 'talking heads' and torsos at the expense of the whole body. It does not create an 'as if' environment as suggested in *Polanski*. Rather it promotes confusion by creating physically separate but perceptually connected spaces.

Viewed in this way live link is not just a substitute for a well-worn method of producing evidence. It is a new way of communicating; a hybrid form of media which relies on a mixture of silicon and genes and generates representations of witnesses that we can hear but not smell, see but not touch. In other words, it is in danger of transforming evidence-giving from the sensual to the sanitised and conveying images through a medium closely associated with fiction. It creates new parameters to relationships which are much more dependent on instant composition and decomposition of subject. The experience of presence and absence presents a cognitive contradiction between sensation and thought which, it is suggested, may leave us confused about what matters in the courtroom and why.

Secondly, Radul (2007) also reminds us that film critics would be keen to instruct lawyers that aspirations to simulate the experience of live evidence are overly ambitious if not impossible. To a large extent this is because the experience is one of technologically mediated intersubjectivities. In short, the transformation of live evidence given in a remote location is far from neutral. Rather, the choice of camera angle and number of cameras create a 'fictionalisation' of what occurs in the separate but related space of the live link video suite. To argue otherwise would be to deny the importance of artistic discussions about camera angles which occur on a daily basis on film sets.

Other characteristics of the live link performance are also worthy of note. The exclusion of spectators from the images of some courtrooms transmitted to the live link witness in their video suite also serves to render impossible a whole series of visual cues about the credibility of what is being said which are often absorbed unconsciously when physically present in the room. As well as inhibiting communication this also engenders a sense of isolation; a public performance from which knowledge of spectators and receipt is excluded. The live link witness is 'in' court but can only see prescribed parts of it. They know they are being observed but cannot see all of their observers.[34] It could be argued that live link produces witnesses to whom a physical or emotional reaction, be it attraction or repulsion, is less likely or less potent. One of the costs which is incurred is a loss of humanity or human connection. The power of legal proceedings to shock, move or anger participants is reduced and the power of the trial as an important social ritual lessened.

Conclusion

It has been argued that public adjudication will always be enriched by the physical presence of participants and that the absence of bodies can serve to impoverish the performance of important public functions and rituals. Elsewhere, the expectation that physical presence is fundamental to dignified proceedings has remained intact. Whilst enthusiasts are keen to introduce technology into the courtroom, there seems to be no parallel call for virtual parliaments, weddings, christenings, bar mitzvahs or funerals. As Katseiris (2009) has argued:

> A society looks to the law court, as well as to other public buildings, to personify the community's state of being with respect to matters of justice. The institution is expected to uphold the law, to demonstrate a certain purity, and to manifest a symbolic weight or an anchorage in an increasingly virtual and dissolving public realm. (p.1)

It would seem then that face-to-face contact is much more likely to confer interaction with meaning and this is surely one of the reasons why mediators continue to encourage us to recognise the particular power of face-to-face interaction in the resolution of disputes. Moreover, when we talk of a litigant's 'day in court' we refer to the basic human need to have one's say in the presence of those who have failed to listen, acknowledge or attach importance to one's narrative in the run-up to trial. In contrast to those who crave efficiency it is concluded that it is the very inconvenience or effort of attending a gathering that should be seen as conferring attendance with importance. Face-to-face contact does not just have procedural value, it has an intrinsic value because it speaks to our political morality.

Notes

1 An earlier version of this chapter appeared as Mulcahy, L. (2008a) 'The unbearable lightness of being – Shifts towards the virtual trial', *Journal of Law and Society*, Vol. 35(4), pp.464–89. I would like to thank the *Journal of Law and Society*, Cardiff University Law School and Blackwell Publishing for their kind permission to reproduce the text.

The dematerialisation of the courthouse 179

2 Several pieces of legislation now allow evidence to be given in this way and the introduction of such provisions has been fuelled by two arguments which need to be clearly distinguished. See, e.g. Police and Justice Act 2006, Part 3A; Criminal Justice Act 2003, Part 8, ss. 51–6; Youth Justice and Criminal Evidence Act 1999, ss. 16–22, 24 and 33A; Home Office Circular 48/2004; Criminal Justice Act 1988, s. 32; Criminal Procedure Rules 30.1; CPR 2.3. See further *Halsbury's Laws of England*, Vol. 11(3) (2006 reissue), paras 1050–7, at 1415.

3 So, e.g. the Court Service launched the Courtroom of the Future project in mid-1998, a Judicial Technology Board has been set up and is chaired by the Judge in charge of modernisation (see further online. Available HTTP: <www.judiciary.gov.uk>) and a Criminal Justice Information Technology group was formed in 2001. Under the JUDITH project, a number of judges have been supplied with personal computers linked to the judicial communications network known as FELIX. A number of pilot schemes have been set up to test the use of particular facilities (see, e.g. online. Available HTTP <http://news.bbc.co.uk/1/hi/uk/395868.stm>). See Bello (1999).

4 Lord Justice Brooke (2004) has argued that the land whose industrial evolution changed the world in the eighteenth and nineteenth centuries has been relatively slow to embrace technology into its legal and judicial culture. Many of the initiatives launched by the Department for Constitutional Affairs (DCA) (now Ministry of Justice) have been focused on installing basic 'plumbing' such as compatible networked PCs, secure intranet sites and email networks for judges, administrators and lawyers. It is, however, clear that modern court-based case-handling systems of the type that are increasingly common overseas are unlikely to be achievable in the usual government spending round. As a result, increasing emphasis is being placed on the provision of IT support through PFI funding.

5 See, e.g. Brooke 2004.

6 In order to understand some of the background behind these measures see Home Office (1998). In *R (on the application of D)* v *Camberwell Green Youth Court* (2005) reference is also made to the fact that there are common law precedents for such provisions. See, e.g. *Smellie* (1919) in which the Court of Criminal Appeal held that the judge could remove the accused from the sight of the witness if the witness was intimidated. For an evaluation of the effectiveness of such procedures, see Hamlyn *et al* (2004). It is worth noting that the Act has been labelled 'schizophrenic' which leads to doctrinal incoherence. See Cooper (2005).

7 In Home Office (1998) it was estimated that around 7–9% of witnesses in the criminal justice system are 'vulnerable'. See also Burton *et al* (2006).

8 See Baroness Hale's judgment in *R (on the application of D)* v *Camberwell Green Youth Court* (2005) in which she makes clear that there is no question, as there was under the old law, of the court striking a balance between the rights of the defendant and those of children. On vulnerable witnesses more generally see Bereton (1997); Birch (2000); Birch and Rees (2001); Burton *et al* (2006); Ellison (1998); Office for Criminal Justice Reform (2006).

9 Omerod (2007) has argued that the rise in gun crime in Britain has led to a widespread refusal of witnesses to co-operate and has created a considerable problem for criminal trials. He goes so far as to suggest that this is challenging the rule of law itself.

10 For some of the literature on these measures see Wade *et al* (1998); Plotnikoff and Woolfson (2008); Cooper (2005).

11 See, in particular, CPR PD 32, Annex 3, para. 2.

12 *Black* v *Pastouna* (2005) at para. 15.

13 See online. Available HTTP: <http://www.justice.gov.uk/news/newsrelease120509a. htm>; <http://www.justice.gov.uk/news/newsrelease300709a.htm>; <http://frontline. cjsonline.gov.uk/_includes/downloads/guidance/cjs-reform/efficiency-and-effectiveness// 20081201_A_Guide_to_Virtual_Courts_in_London_DLk_Leaflet_v1.pdf> (accessed February 2010).

180 Legal architecture: Justice, due process and the place of law

14 This was in the case of *Henderson* v *SBS Realisations* (1992).
15 *Polanski* v *Conde Nast Publications Ltd* CA (2003) at para 60.
16 *Polanski* v *Conde Nast Publications Ltd* HL (2005) in which Lord Nicholls, Lord Slynn, Lord Hope and Lord Carswell all repeated this argument with approval.
17 The defendant's fundamental 'right to confrontation' or altercation has a long history in the UK and the fight to establish and preserve it was particularly fierce in a series of treason trials in the Tudor and Stuart period. See in particular Roberts and Zuckerman (2004), Friedman (1998). See, e.g. *Rex* v *Vipont and others* (1761), p. 767; *King* v *Wilson*. It appears to be a less important aspect of the jurisprudence of the trial in the UK today than in the US where the right to confrontation is enshrined in the Sixth amendment. See further Kirst (2003).
18 See also *Kostovski* v *Netherlands* (1990), *Unterpertinger* v *Austria* (1991), *Windisch* v *Austria* (1991), *Delta* v *France* (1993), Kirst (2003).
19 McEwan (1998) and Roberts *et al* (2005).
20 See, e.g. CPR PD 32, Annex 3, para. 2.
21 See further online. Available HTTP: <http://www.courtroom21.net>.
22 For the purposes of the project a 'high tech' courtroom is one which has several features including real-time transcription of oral evidence with a facility for the judge and lawyers to mark and annotate such text as the evidence is given; facilities for the presentation of evidence and other material in a visual, possibly computer-generated, form; and the capacity to receive evidence or 'allow attendance' by remote video link (West-Knights 2005).
23 At a more parochial level, the material is commonly used to anchor court buildings in their local community. The existence of the Middlesex coat of arms and war memorials to the Middlesex Regiment which adorn the exterior and interior of Middlesex Guildhall, the display of the coat of arms of the local law society and commercial associations at Manchester High Court, and the statue of Thomas Telford which stands outside Telford County court make clear that these are not courthouses that could have sensibly been built elsewhere. Instead, they celebrate the connection between particular communities and legal process and link the invisible reality of law to local soil.
24 CPR PD 32, Annex 3, para 14. In Northern Ireland the negative symbolic significance of the coat of arms has been recognised in recent legislation which allows that it should no longer be displayed in newly built courthouses. The rationale behind this is that it has become associated with partial justice in the province.
25 CPR PD 32, Annex 3. For guidance in the criminal sphere, see Crown Prosecution Service, online. Available HTTP: <http://www.cps.gov.uk/legal/section13/chapter_m.html>.
26 See for instance online. Available HTTP: <http://www.newsshopper.co.uk/video_audio/videonews/88252/> or <http://news.bbc.co.uk/1/hi/england/london/8070279.stm>. (Accessed February 2010).
27 Online. Available HTTP: <http://www.justice.gov.uk/news/newsrelease300709a.htm> (Accessed February 2010).
28 See further the Police and Justice Act 2006.
29 *Polanski* v *Conde Nast Publications Ltd* CA (2003).
30 See, in particular, Temkin (2000); Carlen (1976). Cooper (2005) suggests that for children the formal courtroom is an 'arcane and mysterious corner of the adult world' and that this works against the effective delivery of testimony.
31 The quality of live link transmissions in contemporary trials will not be explored in any depth. This is certainly an important issue but there seems little doubt that the quality of transmission will improve and that in time images of witnesses giving evidence from remote locations will appear as sophisticated 3D holograms within the body of the court.
32 *Polanski* v *Conde Nast Publications Ltd* (2005).

33 See, e.g. Blumenthal 1993; Burton *et al.* 2006; Stone 1991; Costigan and Thomas 2000. These authors argue that there is ample research by psychologists to show that confidence is not an indication of reliability, inconsistency is not synonymous with inaccuracy, and that witnesses often appear nervous because stressed. Stone (1991) argues that there is no known physiological connection between the brain processes of a lying person and bodily or vocal signs. He also claims that psychological research confirms that there are no specific signs of lying.

34 Video link screens within the court are commonly positioned so that spectators in the public gallery can see the images. However, the live link witness is not able to see the spectators.

Bibliography

Abel, R. (2003) *English Lawyers Between Market and State, The Politics of Professionalism*, Oxford: Oxford University Press.

Archer, P. (1992) 'Video First at County Court', Press Association 13 May, available on www.lexisnexis.co.uk, last accessed 20/03/07.

Arnheim, W. and Watterson, J. (1966) 'Inside and Outside in Architecture', *The Journal of Aesthetics and Art Criticism*, Vol. 25, No. 1, pp.3–15.

Arnold, R. and Merritt, G. (1999) 'By the consent of the governed' in Giles, R., and Snyder, R., *Covering the Courts: Free Press, Fair Trials and Journalistic Performance*, New Brunswick: Transaction Publishers.

Arthurs, H. (1984) 'Special Courts, Special Law: Legal Pluralism in c19th England', pp.380–411 in Rubin, G. and Sugarman, D., *Law, Economy and Society 1750–1914: Essays in the History of English Law*, London: Professional Books Ltd.

Aslet, C. (1982) 'A Palace for many courts: Centenary of the Royal Courts of Justice', *Country Life*, November 11, pp.1462–3.

Attorney General's Office, http://www.attorneygeneral.gov.uk/AboutUs/Pages/History.aspx Last accessed April 2010.

Bailey, J. (2001) 'Voices in court: lawyers or litigants?' *Historical Research*, Vol. 74, no. 186, pp.392–408.

Barker, H. (2000) *Newspapers, Politics and English Society 1695–1855*, London: Longman.

Barker, J. (1964) 'Court House Ampthill' 7th February, *The Builder*, pp.283–5.

Barrister (1978) 42 *Journal Criminal Law*, 3.

Belfast Newsletter, The (1884) 'Belfast Recorders Court' 13 February, Issue 21394.

Belfast Newsletter, The 'Petty Sessions' 3 September 1898, Issue 25922.

Bell, H. (1953) *An introduction to the History and Records of the Court of Wards and Liveries*, Cambridge: Cambridge University Press.

Bello, R. (1999) 'Towards the Virtual Courtroom', www.practicallaw.com/6-101-0477, last accessed 28/07/08.

Bellott, H. (1922) 'Some early courts and the English Bar', 38 *Law Quarterly Review*, pp.168–84.

Bentham, J. (1827) *Rationale of Judicial Evidence*, ed. by Stuart Mill, J. Book II, ch x, London: Hunt and Clarke.

Bereton, B. (1997) 'How Different are Rape Trials? A Comparison of the Cross Examination of Complainants in Rape and Assault Trials', 37(2) *British Journal of Criminology* 242–61.

Better Public Building Initiative (2008) http://www.betterpublicbuilding.org.uk/finalists/2008/ Last accessed April 2010.

184 Bibliography

Birch, D. (2000) 'A Better Deal for Vulnerable Witnesses?' *Criminal Law Review* 223.

Birch, D. and Rees, T. (2001) 'Case Comment: Trial – Child Complainant, Cross Examination by Judge', *Criminal Law Review* 587.

Birmingham Daily Post (1891a) 'The Royal Visit', Monday July 20, Issue 10319.

Birmingham Daily Post, (1891) 'News of the day' Tuesday July 28, 1891, Issue 10326.

Birmingham Daily Post, December 25, 1893, Issue 11081.

Black, M. (2009) Representations of justice, A photographic Essay, JOSCCI, number one, www.uow.edu.au/arts/jiscci. Last accessed November 2009.

Blackstone, W. (1765–9) *The Commentaries of Sir William Blackstone, Knight, on the Laws and Constitution of England*, Reproduced (2010) Chicago: American Bar Association.

Blatcher, M. (1978) *The Court of King's Bench 1450–1550*, London: The Athlone Press.

Blomley, N., Delaney, D., and Ford, R. (eds) (2001) *The Legal Geographies Reader*, Oxford: Blackwell.

Blumenthal, J. (1993) 'A Wipe of the Hands, a Lick of the Lips: the Validity of Demeanour Evidence in Assessing Witness Credibility', *Nebraska Law Review* 1157.

Boyce, G. (1978) 'The fourth estate: the reappraisal of a concept' in *Newspaper History from the 17th century to the present day*, eds George Boyce, James Curran and Pauline Wingate, London: Constable.

Brawn, G. (2009) 'The changing face of justice: The architecture of the Australian courthouse', *Architecture Australia*, Vol. 98(5), pp.39–42.

Briggs, A. (1950) 'Social structure and politics in Birmingham and Lyons (1825–1848), *The British Journal of Sociology*, Vol. 1 no. 1 pp.67–80.

Briggs, A. (1956) 'Middle class consciousness in English Politics 1780–1846', *Past and Present*, no. 9 pp.65–74.

Briggs, A. (1993), *Victorian Cities*, Berkeley: University of California Press.

British History online (2009a). From: 'The city and liberty of Rochester: General history and description', The History and Topographical Survey of the County of Kent: Volume 4 (1798), pp.45–86. URL: http://www.british-history.ac.uk/report.aspx?compid=53798 &strquery=Guildhall. Date accessed: 21 January 2009.

British History online (2009b) The hundred of Cashio: Introduction', in *A History of the County of Hertford*: 2:319–322 (1908) http://www.british-history.ac.uk Date accessed: 1 April 2009.

British History online (2009c) 'The hundred of Mainsbridge', in *A History of the County of Hampshire* 3:462 (1908) http://www.british-history.ac.uk Date accessed: 1 April 2009.

British History online (2009d) 'The hundred of Eastbourne: Introduction', in *A History of the County of Sussex*: 4:40 (1953) http://www.british-history.ac.uk Date accessed: 1 April 2009.

British History online (2009e) 'The city of St Albans: The borough', in *A History of the County of Hertford* 2:477–483 (1908) http://www.british-history.ac.uk Date accessed: 1 April 2009.

British History online (2009f) 'Bardsey – Barkway', in *A Topographical Dictionary of England* (1848) 141–145 http://www.british-history.ac.uk Date accessed: 1 April 2009.

British History online (2009g) 'Parishes: St Stephen's', in *A History of the County of Hertford*: 2:424–432 (1908) http://www.british-history.ac.uk Date accessed: 1 April 2009.

British History online (2009h) 'Faircross hundred: Introduction', in *A History of the County of Berkshire* 4:38–39 (1924) http://www.british-history.ac.uk Date accessed: 1 April 2009.

British History online (2009i) 'Hundred of Shropham: Atleburgh', in *An Essay towards a Topographical History of the County of Norfolk*: 1:501–541 (1805). http://www.british-history.ac.uk Date accessed: 1 April 2009.

British History online (2009j) 'Norton Ferris Hundred', *A History of the County of Somerset*: Volume 7: Bruton, Horethorne and Norton Ferris Hundreds (1999), pp.161–163. URL: http://www.british-history.ac.uk Date accessed: 1 April 2009.

British History online (2009k) Gloucester Boothall see 'Gloucester: Public buildings', (1998) in *A History of the County of Gloucester* Volume 4: The City of Gloucester: 248–251. Online. Available http://www.british-history.ac.uk Date accessed 1 April 2009.

Brodie, A., Winter, G., and Porter, S. (2001) *The Law Court 1800–2000: Development in form and function*, London: English Heritage. Unpublished.

Brodie, A., Davies, J. and Croom, J. (1999) *Behind Bars: the High Architecture of England's Prisons*, London: Royal Commission on the Historical Monuments of England.

Brooke, Lord Justice (2004) 'The legal and policy implications of courtroom technology: the emerging English experience', 13 February, http://www.judiciary.gov.uk/publications_media/speeches/2004/ljb130204.htm, last accessed 28/07/08.

Brookes, C. and Lobban, M. (1997) *Communities and Courts in Britain*, London: The Hambledon Press.

Brooks, C. (1986) *Pettyfoggers and vipers of the commonwealth: The lower branch of the legal profession in early modern England*, Cambridge: Cambridge University Press.

Brophy, J. and Roberts, C. (2009) *'Openness and transparency' in family courts: what the experience of other countries tells us about reform in England and Wales*, Family Policy Briefing 5, Department of Social Policy and Social Work, Oxford. See also www.spsw.ox.ac.uk (last accessed October 2009).

Brown, F. (1977) 'Preface' p. iii in *Magistrates' Courthouses: Design Study 1977*, London: Home Office and the Greater London Council.

Brown, J. (1980a) 'Design for Law and Order: A general survey', *Architects Journal*, June 1980 pp.1191–2.

Brown, J. (1980b) 'Design for Law and Order: Two case studies', *Architects Journal*, June 1980 p1235.

Brown, J. (1987) 'The Law of Architecture: Glasgow Sheriffs Courts' *Architects Journal*, 11 February pp.29–32.

Brownlee, D. (1983), 'To agree would be to commit an act of artistic suicide, the revision of the design for the law courts', *The Journal of the Society for Architectural Historians*, Vol. 42 no. 2, May 1983 pp.168–188.

Brownlee, D. (1984) The *Law Courts – The Architecture of George Edmund Street*, New York: The Architectural Foundation.

Builder, The (1848) Vol. VI, no 304, Nov 25 p.577.

Builder, The (1854), 'Where shall the new law courts be built?', March 11, Vol. XII, p.128.

Builder, The (1859) April 23, Vol. XVII, p.289.

Builder, The (1862) 'Proposed New Courts of Justice', April 19th, p.276. February 3, p.138.

Builder, The (1882), 'Courts and courts: A reverie', December 9 p.746.

Builder, The (1883) 'In and around the Royal Courts of Justice', February 3, pp.138–9.

Builder, The (1898) 'Lifts at the law courts', June 4, p.540.

Builder, The (1899) 'Improvements at the Law Courts', March 25.

Builder, The (1902) 'Accommodation at the Law Courts', May 17, p.491.

Builder, The (1911) 'Accommodation for witnesses at Law Courts', January 27, p.99.

Builder, The (1957) 'Law Courts, Slough', October 18 pp.665–70.

Builder, The (1963) 'Plymouth Law Courts', October 4, pp.667–9.

Builder, The (1978) 'Chester', 3 August, p.818.

Building News (1858), 'Manchester Assize Courts', 6 May, pp.421–2.

Building News (1859) 'Manchester Assize Court Competition', 29 April, pp.393–4.

Bürklin, T., Limbach, J. and Wilkens, M. (2004) 'With a touch of Internationality and Modernity', *The Federal Constitutional Court of Germany: Architecture and Jurisdiction*, Basel: Burkhäuser Publishers.

Burroughs, C. (2000) 'Spaces of Arbitration and the organisation of space in late medieval Italian Cities', in *Medieval Practices of Space*, (eds) Hanawalt and Kobialka, Minneapolis: University of Minnesota Press.

Burton, M., Evans, R. and Sanders, A. (2006a) 'Implementing Special Measures for Vulnerable and Intimidated Witnesses: the Problem of Identification', *Criminal Law Review* 229.

Burton, M., Evans, R. and Sanders, A. (2006b) *An Evaluation of the Use of Special Measures for Vulnerable and Intimidated Witnesses*, Home Office Findings 270.

Burton, M., Evans, R. and Sanders, A. (2006c) 'Are Special Measures for Vulnerable and Intimidated Witnesses Working?', Evidence from the Criminal Justice Agencies' Home Office Online Report 01/06.

Burton, M., Evans, R. and Sanders, A. (2007) 'Vulnerable and Intimidated Witnesses and the Adversarial Process in England and Wales' *International Journal of Evidence and Proof* 1–23.

Cannadine, D. (1977) 'Victorian cities: How different?' 2 *Social History* 457. This article is also reproduced in R. Morris and R. Rodger (1993) (eds), *The Victorian City: A reader in British Urban History*, London: Longman.

Carlen, P. (1974) 'Remedial Routines for the Maintenance of Control in Magistrates' Courts', *British Journal of Law and Society*, Vol. 1, No. 2, pp.101–117.

Carlen, P. (1976) *Magistrates' Justice*, London: Martin Robertson.

Central Criminal Court (2010) http://www.oldbaileyonline.org/static/The-old-bailey.jsp Last accessed April 2010.

Chalklin, C. (1998) *English Counties and Public Building 1650–1830*, London and Rio Grande: The Hambledon Press.

Chalklin, C. (ed) (1984) *New Maidstone Gaol Order Book 1805–23*, Kent: Kent Archaeological Society.

Chandler, J. and Dagnall, H. (1981) *The Newspaper and Almanac Stamps of Great Britain*, The Newspapers Handbook, Britain and Ireland, Essex: Philatelic Publications Ltd.

Chase, O. (2005) *Culture and Ritual*, New York: New York University Press.

Cioni, M. (1985) *Women and law in Elizabethan England with particular reference to the court of Chancery*, New York: Garland Press.

The City of Gloucester (1988), 'Gloucester: Public buildings', A History of the County of Gloucester: Volume 4: pp.248–251. URL: http://www.british-history.ac.uk/report.aspx?compid=42305&strquery=Gloucester shire hall Last accessed: January 2009.

Chorley, Lord, (1970), 'The Report of the Royal Commission on Assizes and Quarter Sessions', *Modern Law Review*, Vol. 33, No. 2 pp.184–90.

Chyutin Architects (2008), *Architecture for Justice: The Haifa Court House*, Jerusalem: Hebrew University Magnes Press.

Clarke, P. and Martin, L. (1996) *The History of Maidstone: the making of a modern county town*, Maidstone: Alan Sutton publishers.

Clegg, C. (1997) *Press Censorship in Elizabethan England*, Cambridge: Cambridge University Press.

Cochlin, R. (2002) *Get into Citizenship: Crime and legal awareness*, London: PtP Publishing.

Cockburn, J. (1972) *A History of English Assizes*, Cambridge: Cambridge University Press.

Colinvaux, R. (1967) 'Journalism and the law' in *Inside Journalism*, (ed) Bennett-England, R. London: Robert Owen Limited.

Collingwood, F. (1967) 'Robert Smirke', *Building* 14 April p.92.

Collins, P. (2002–2003) 'South Eastern Circuit' in *County Court Annual Report*, London: Department for Constitutional Affairs.

Colvin, H. (1977) *The History of the King's Works*, Vol. V, 1660–1782, London: Her Majesty's Stationery Office Books.

Concise Oxford English Dictionary (2008) Twelfth edition. Ed. Catherine Soanes and Angus Stevenson, Oxford: Oxford University Press.

Construction (1992a) 'Northampton Crown and County Courts', 85 pp.39–41.

Construction (1992b) 'Newport Crown Court', 85, pp.54–5.

Cooper, D. (2005) 'Pigot Unfulfilled: Video Recorded Cross Examination under s. 28 of the Youth and Criminal Evidence Act 1999' *Criminal Law Review* 456.

Corner, G. (1863) 'Observations on four illuminations', *Archaeologia*, Vol. xxxix, pp.357–72.

Corner, G. (1865) *Observations on four Illuminations Representing the courts of chancery, King's Bench, Common pleas and Exchequer at Westminster From a manuscript of the time of Henry VI, Communicated to the society of antiquaries*, Nichols and Sons, Westminster 1865.

Cornish, W. and Clark, G. (1989) *Law and Society in England 1750–1950*, London: Sweet and Maxwell.

Costigan, R. and Thomas, P. (2000) 'Anonymous Witnesses' 51(2) *Northern Ireland Legal Quarterly* 326–58.

Country Life (1922) 'The Costume of the Law', January 28, pp.108–109.

County Court Chronicle (1846) issue one, p.11.

County Court Chronicle (1847–1849) Volume one, no. 4, Law Times, London.

County Court Chronicle (1854) October p.236.

County Court Chronicle (1856) 'Intelligence'. Sept, p.174.

County Court Chronicle (1857) Oct 1 pp.178–9 'County court returns'.

County Court Chronicle (1860) 'Statistics of the County Courts' pp.78–9, Aug 1.

Cranfield, G. (1962) *The Development of the Provincial Newspaper 1700–1760*, Westport, Connecticut: Greenwood Press Publishers.

Cranfield, G. (1978) *The Press and Society*, London: Longman.

Cretney, S.M. (1998) 'Disgusted. Buckingham Palace . . . Divorce, Indecency and the Press 1926' in *Law, Law Reform and the Family*, Oxford: Clarendon Press, pp.91–114.

Crook, J. (1967a) 'Architect of the Rectangular: A reassessment of Robert Smirke', *Country Life*, April 13, pp.846–8.

Crook, J. (1967b) 'Sir Robert Smirke, a century florilegium', *Architectural Review*, September pp.208–210.

Cunningham, C. and Waterhouse, P. (1992) *Alfred Waterhouse 1830–1905: Biography of a Practice*, Oxford: Clarendon Press, pp.31–5; 212; pls 33–6, 174.

Cunningham, C. (1985) *Building for the Victorians*, Cambridge: Cambridge University Press.

188   Bibliography

Curran, J. and Seaton, J. (1997) *Power without responsibility: The Press and broadcasting in Britannia* (5th edition), London: Routledge.

Daily News (1858) 'The Queen's Visit to Leeds', Wednesday September 8, Issue 3843.

Damaska, M. (1986) *The Faces of Justice and State Authority*, New Haven: Yale University Press.

—— (1997) *Evidence Law Adrift*, London: Yale University Press.

Davidson, E. (2000) *The Lawyer*, 22 September 1998, p.6.

Davies, G. (1999) 'The Impact of Television on the Presentation and Reception of Children's Testimony', 22 *International Journal of Law and Psychiatry* 241–56.

Davis, J, (1984) 'New Law Courts, Maidstone', *Construction*, 48, pp.19–21.

Denton, J. (2009) 'Courts in the UK', *Architecture Australia*, Vol. 98(5), pp.53–7.

Department for Constitutional Affairs (2000) *Civil Justice 2000: A Vision of the Civil Justice System in the Information Age*, London: Department for Constitutional Affairs.

Department for Constitutional Affairs (2003a) *Magistrates and Schools: A Guide to court visits*, London: Department for Constitutional Affairs.

Department for Constitutional Affairs (2003b) *A Guide to court visits in the crown and county court*, London: Department for Constitutional Affairs.

Department for Constitutional Affairs (2005) 'Public private partnerships and the PFI', www.dca.gov.uk/courtbuild. Last accessed August 2005.

Department for Constitutional Affairs (2006) *Confidence and confidentiality: Improving transparency and privacy in family courts*, London: Department for Constitutional Affairs.

Derby Mercury, The (1864) Wednesday August 3, p.6 col 4.

Dicey, A. (1914) *Lectures on the relation between law and public opinion in England during the nineteenth century* (2nd Edition), London: Macmillan and Co.

Doerksen, L. (1989–90) 'Out of the dock and into the bar: An examination of the history and use of the prisoner's dock', 32 *Criminal Law Quarterly*, p.478.

Douzinas, C. (1999) 'Prosopon and Antiprosopon: Prolegomena for a legal Iconology' in Douzinas, C. and Nead, L. (eds), *Law and the image – the authority of art and the aesthetics of law*, Chicago: University of Chicago Press.

Douzinas, C. and Warrington, R. with McVeigh, S. (1991) 'Interlude and supplement: A written lecture on writing in trials', pp.151–7 in *Postmodern Jurisprudence: the Law of text in the texts of law*, London: Routledge.

Dovey, K. and Fitzgerald, J. (2010) 'Open court: Transparency and Legitimation in the Courthouse', pp.125–38 in *Becoming Places: Urbanism/Architecture/Identity/Power*, London: Routledge.

Duff, A., Farmer, L., Marshall, S. and Tadros, V. (2007) *The trial on trial: towards a normative theory of the trial, volume three*, Oxford: Hart Publishing.

Edwards, A. (1993) 'Courting Design', *Architecture*, February pp.89–90.

Edwards, Trystan (1914) 'Modern Architects III – Sir Robert Smirke', *The Architects and Builders Journal*, May 27, pp.369–71.

Ellison, L. (1998) 'Cross Examination in Rape Trials', *Criminal Law Review* 605–15.

Evans, D. (1999) 'Theatre of deferral: The image of the Law and the architecture of the inns of court', *Law and Critique*, Vol. 10(1), pp.1–25.

Fairweather, L. and McConville, S. (eds) (2000), *Prison Architecture: Policy, Design and Experience*, Oxford: Architectural Press.

Ferguson, R. (1984) 'Commercial Expectations and the Guarantee of the Law: Sales transactions in mid 19th century England', pp.192–208 in Rubin, G.R., and Sugarman, David, *Law, Economy and Society 1750–1914: Essays in the History of English Law*, London: Professional Books Ltd.

Finn, J. (1996) 'Political or Professional Honours', *Australian Journal of Legal History*, Vol. 2, issue 1, pp.61–78.

Fischer-Taylor, K. (1993) *In the Theater of criminal justice: The Palais de Justice in Second Empire Paris*, Princeton, NJ: Princeton University Press.

Florence, S. (2003) ' "We Do Not Have It, and We Do Not Want It": Women, Power, and Convent Reform', *The Sixteenth Century Journal*, Vol. 34, No. 3, pp.677–700.

Foster, H. (1961–62), 'Common Law Divorce', 46 *Minn. L. Rev* 43.

Foucault, M. (1977) *Discipline and Punishment – the birth of the prison*, London: Penguin. Translated by Alan Sheridan.

Foucault, M. (1984) 'Space, Knowledge and Power' in Rabinow, P. (ed), *The Foucault Reader*, London: Penguin.

Foucault, M. (2008) *Spectacle of the scaffold*, London: Penguin.

Freeman's Journal and Daily Commercial Advertiser (1844), 'Court of Queen's Bench', 12 January.

Freeman's Journal and Daily Commercial Advertiser (1848), 'The Official version of the unpublished proceedings at Green Street', 22 December.

Freschi, F. (2007) 'Postapartheid Public and the Politics of Ornament: Nationalism, Identity, and the Rhetoric of Community in the Decorative Program of the New Constitutional Court, Johannesburg', *AfricaTODAY*, Vol. 54(2), pp.178–94.

Friedman, R. (1998) 'Thoughts from Across the Water on Hearsay Confrontation', *Criminal Law Review* 697.

Frishman, M. (1994) 'Islam and the form of the mosque', pp.17–42 in *The Mosque: History, Architectural development and regional diversity*, ed by Martin Frishman and Hasan-Uddin Khan, London: Thames and Hudson.

Fulford, R. (1998) 'Correct architecture and the architecture of corrections', *Globe and Mail*, September 19. See: http://www.robertfulford.com/Courthouses.html

Galanter, M.M. (2004) 'The vanishing trial: an examination of trials and related matters in federal and state courts', *Journal of Empirical Studies*, Vol 1, Issue 3, pp.459–570.

Gale, P. (2006) *Pride of place, the story of Abingdon's County Hall*, Oxford: Trafford publishing.

Garrard, J. (1983) *Leadership and Power in Victorian Industrial Towns 1830–80*, Manchester: Manchester University Press.

Genn, H. (2009) *Judging Civil Justice, The Hamlyn Lectures 2008*, Cambridge: Cambridge University Press.

Gerhold, D. (1999) *Westminster Hall: Nine hundred years of history*, London: James and James Publishers Ltd.

Gibb, F. (2010) 'Sky begins push for TV cameras in court', January 13, Timesonline. http://business.timesonline.co.uk/tol/business/law/article6986541.ece Last accessed February 2010.

Girouard, M. (1980) *Life in the English Country House: A social and architectural history*, London: Penguin Books.

Girouard, M. (1990) *The English Town*, London: Guild Publishing.

Glancey, J. (2006) 'Antwerp's soaring new law court', The Guardian, Monday 10 April 2006. <http://www.guardian.co.uk/artanddesign/2006/apr/10/architecture

190 Bibliography

Gold, J. (1997) *The experience of modernism: Modern architects and the future city 1923–53*, London: E&N Spon.

Goodhart-Rendel, H. (1963) 'Victorian Public Buildings', pp.85–103 in Peter Ferriday (ed), *Victorian Architecture*, London: Jonathan Cape Publishers.

Goodrich, P. (1987) *Legal Discourse: Studies in Linguistics, Rhetoric and Legal Analysis*, London: Macmillan.

Graham, C. (2003) *Ordering Law – the architecture and social history of the English law court to 1914*, Aldershot: Ashgate Publishing.

Graham, C. (2004) 'The history of law court architecture in England and Wales; The institutionalization of the law', pp.36–47 in SAVE Britain's Heritage, *Silence in court: The future of the UK's historic law courts*, London: SAVE Britain's Heritage.

Grahame, S. (2003) 'Highlights on Leeds Town Hall', Leeds Council (2009) http://www.leeds.gov.uk/discover/discovery.asp?page=200335_519878567. Last accessed January 2009.

Grant, E. (2009) 'Port Augusta Courts', *Architecture Australia*, Sept/Oct 2009, pp.86–90.

Green, A. and De Nevi, D. (1989) *An architecture for democracy: Frank Lloyd Wright – The Marin County Civic Centre*, California: Horizon Press.

Greene, F. (2006) 'The image of the courthouse' in *Celebrating the courthouse: A Guide for Architects, their clients and the public*, ed by Steven Flanders, New York: Norton and Co.

Greenhouse, L. (1996) Telling the Courts Story: Justice and Journalism in the Supreme Court, 105 *Yale Law Journal*, p.1537.

Gretton, T. (1980) *Murders and moralities, English catchpenny prints 1800–1860*, London: British Museum Publications.

Griffiths, A. and Kandel, R. (2009) 'The Myth of the Transparent Table: Reconstructing Space and Legal Interventions in Scottish Children's Hearings' in *Spatializing Law: An Anthropological Geography of Law in Society*, edited by Franz von Benda-Beckmann, Keebet von Benda-Beckmann and Anne Griffiths, Farnham: Ashgate.

Gross, C. (1906) 'The Court of Piepowder', *The Quarterly Journal of Economics*, Vol. 20, No. 2, pp.231–49.

Grossman, J. (2002) *The Art of Alibi: English Law Courts and the Novel*, Baltimore: The Johns Hopkins University Press.

Haldar, P. (1999) 'The Function of the ornament in quintillian, Alberti and Court Architecture' in Douzinas, C., and Nead, L. (eds), *Law and the image – the authority of art and the aesthetics of law*, Chicago: University of Chicago Press.

Hamlyn, B., Phelps, A. and Sattar, G. (2004) *Key Findings from the Survey of Vulnerable and Intimidated Witnesses 2001/01 and 2003*.

Hanbury, H. and Yardley, D. (1979) *English Courts of Law*, Oxford: Oxford University Press.

Haour, A. (2005) 'Power and Permanence in Precolonial Africa: A Case Study from the Central Sahel', *World Archaeology*, Vol. 37, No. 4, Debates in 'World Archaeology' (Dec.), pp.552–65.

Harding, A. (1973) *The Courts of Medieval England*, London: George Allen and Unwin Ltd.

Harris, A. (1959) 'New Courthouse at Corby', *The Builder*, December 11, pp.843–5.

Harvey, D. (1996) 'The currency of time-space', chapter 10 in *Justice, Nature and the Geography of Difference*, Oxford: Blackwell.

Haskins, G. (1948–49) 'Development of Common Law Dowries', 62 *Harvard Law Review* 46.

Bibliography 191

Hasted, E. (1798) *The History and Topographical Survey of the County of Kent*, Volume 4 by Edward Hasted, pp.324–53. See http://www.british-history.ac.uk/report.aspx?compid =53810&strquery=Boxley Date accessed 20 July 2009.

Hastings, M. (1947) *The Court of Common Pleas in Fifteenth Century England: A study of legal administration and procedure*, New York: Cornell University Press.

Hawkins, G. (1982) *Stonehenge Decoded*, London: HarperCollins.

Hennesey, Chief Justice (1984) 'Foreword', in *Courthouses of the Commonwealth*, ed. Brink, R. p. vii, photographs by G. Peet and G. Keller. Amherst: University of Massachusetts Press.

Henocq, T. (1999) 'Court Building Design: A review of the programme', *Construction*, 69 pp.5–8.

Henriques, E. (1972) 'The rise and decline of the separate system of prison discipline', *Past and Present*, 54(1), 61–93.

Her Majesty's Courts Service (2009) *Planning for the future of the magistrates' courts service in London*, London: Her Majesty's Courts Service.

Her Majesty's Courts Service (2010) *Court Standards and Design Guide*, London: Her Majesty's Courts Service.

Herber, M. (1999) *Legal London – A pictorial history*, Chichester: Phillimore and Co.

Herman, S. (2006) *The right to a speedy and public trial*, Santa Barbara: Greenwood Press.

Hidden Horsham, http://www.hiddenhorsham.co.uk/35/oldtownhall.htm Last accessed July 2009.

History of the County of Warwick (1945) 'The hundreds of Warwickshire': Volume 3: Barlichway hundred pp.1–4. URL: http://www.british-history.ac.uk/report.aspx? compid=56972&strquery=moot hill. Date accessed 1 April 2009.

Hobsbawm, E. (1962) *The Age of Revolution, Europe 1789–1848*, London: Weidenfeld and Nicolson.

Hobsbawm, E. (1975) *The Age of Capital 1848–75*, London: Weidenfeld and Nicolson Ltd.

Hockings, J. (2009) 'Brisbane Supreme and District Court', *Architecture Australia*, Sept/ Oct 2009, pp.65–7.

Holdsworth, W. (1903) *A History of English Law, volume one*, London: Methuen and Co.

Home Office (1998) *Speaking up for Justice: Report of the Interdepartmental Working Group on the Treatment of Vulnerable or Intimidated Witnesses in the Criminal Justice System*.

Home Office and the Greater London Council (1977) *Magistrates' Courthouses: Design Study 1977*.

Homer (1950) *The Iliad*, trans. Rieu, E., London: Penguin Classics.

Hoon, G. (1999) 'Crime, Criminal Justice and the Internet' www.dca.gov.uk/ speeches/1999/1-4-99.htm, last accessed 25/10/05.

Howard, J. (1777) *The State of Prisons in England and Wales. Vol 1* (4th edition, 1971), London: Howard League for Penal Reform.

Howe, S. (1846) 'An essay on separate and congregate systems of prison discipline', *Prison Discipline Pamphlet*, Vol. 5, no. 4, Boston: Ticknor.

Howard League for Penal Reform (1976) *No brief for the dock, Report of the Howard League Working Party on Custody during trial*, Chair Lady James of Rusholme, Sussex: Barry Rose Publishers.

Hoyano, I. (2007) 'The Child Witness Review: Much Ado About Too Little', *Criminal Law Review* 849.

Hunt, T. (2005) *Building Jerusalem – The rise and fall of the Victorian City*, London: Phoenix.

192 Bibliography

Hurst, H. (1992) Shaping a new order in the court: A sourcebook for juvenile and family court design, Pittsburgh, PA: National Center for juvenile justice.

Illustrated Paper and Illustrated Times, The (1887) 'Birmingham's grand welcome to our Queen', Saturday April 2, p.212 Issue 1348.

Illustrated Police News, The (1867a), 'Youthful depravity and alleged parental neglect', Saturday February 2, Issue 155.

Illustrated Police News, The (1867b), Saturday February 23, Issue 158.

Illustrated Police News, The (1867c), 'Juvenile thieves', Saturday February 23, Issue 158.

Ingleton, R. (2009) The Early Days of Policing in Kent Part 2. http://www.kent-police-museum.co.uk/core_pages/pasttimes_early_days_pt2.shtml Last accessed July 2009.

Jacob, R. (1995–6) 'The historical development of courthouse architecture/La formazione storica dell'architectura giudiziaria, *Zodiac*, 14: 30–43.

Jackson, I. (1972) *The Provincial Press and the Community*, Manchester: Manchester University Press.

Jackson, J. (2005) 'Effect of Human Rights on Evidentiary Processes', 68 *Modern Law Review* 737.

Jaconelli, J. (2003) 'What is a trial?' in *Judicial tribunals in England and Europe 1200–1700, The trial in history, vol. 1* edited by Mulholland, M., and Pullan, B., with Pullan A., Manchester: Manchester University Press.

Jeavons, K. (1992) 'A proud tradition of public service', *Construction* (85) p.31.

Jenner, M. (2000) *Victorian Britain*, London: Weidenfeld and Nicolson.

Jenkins, F. (1963) 'The Victorian architectural profession' in Peter Ferriday (ed), *Victorian Architecture*, London: Jonathan Cape Publishers.

Jones, D. (2009) Historic courts: Between Civic Memory and a Future of Hope, *Architecture Australia*, Vol. 98(5), pp.59–63.

Jones Baker, Doris (ed) (2005) *Hertfordshire in History*, University of Hertfordshire: Hertfordshire Publications.

Jones, Mr. (1871) *London characters and the humorous side of London Life*. Stanley Rivers and Co, London. See also www.victorianweb.org Last accessed April 2010.

Katsieris, P. (2009) 'The Design of the Commonwealth Law Courts Building, Melbourne', *JOSCCI*, number one, www.uow.edu.au/arts/joscci. Last accessed November 2009.

Kerr, M., Forsyth, R. and Plyley, M. (1992) 'Cold water and hot iron: Trial by Ordeal in England', *Journal of Interdisciplinary History*, Vol. 22, No. 4, pp.573–95.

Kinnard, J. (1963) 'GE Street The Law Courts and The Seventies' in Peter Ferriday (ed), *Victorian Architecture*, London: Jonathan Cape Publishers.

Kirke, P. (2009) 'Kalgoorie Courts Project', *Architecture Australia*, Sept/Oct, pp.71–5.

Kirst, R. (2003), 'Hearsay and the Right of confrontation of the ECHR', 21 *Quinnipac Law Review* 777.

Kitchen, S. and Elliot, R. (2001) *Key Findings from the Vulnerable Witness Survey*, Home Office Findings 147.

Kostal, R. (1994) *Law and English railway capitalism, 1825–1875*, Oxford: Clarendon Press.

Kostoff, S. (1995) *A History of Architecture: Rituals and Settings* (second edition), Oxford: Oxford University Press.

Landsman, S. (1980) 'The decline of the adversary system: how the rhetoric of swift and certain justice has affected adjudication in American Courts', *Buffalo Law Review*, Vol. 29, pp.487–530.

Langbein, J. (2003) *The Origins of the Adversary Criminal Trial*, Oxford: Oxford University Press.

Bibliography 193

Larson, M. (1979) *The Rise of Professionalisation: A sociological analysis*, Berkeley: University of California Press.

Lassiter, C. (1996) 'TV or not TV – That is the question', *The Journal of Criminal Law and Criminology*, Vol. 86, no 3, pp.928.

Law Society (1976) Memorandum by the Council of 'The Use of the Dock in criminal courts' appendix, Howard League for Penal Reform *No brief for the dock, Report of the Howard League Working Party on Custody during trial*, Chair Lady James of Rusholme, Sussex: Barry Rose Publishers.

Law Times (1859) 'Plans for the new courts' 4 June, pp.138–9.

Law Times (1862) 'Assize intelligence – home circuit', 9 August, p.520.

Law Times, (1863) 8 August, p.508.

Law Times, (1864a), 'Legal Topics of the Week', 2 July, p.389.

Law Times (1864b) 'Assize Intelligence – Northern Circuit', 6 August, p.449.

Law Times (1864c) 26 November 1864, p.43.

The Lawyer (2009) http://www.thelawyer.com/bar-fails-to-lure-ethnic-minorities-and-women/115290.article

Lederer, F. (1991) 'Courtroom Practice in the 21st Century Trial' www.ncsonline.org/D_KIS/TechInfo/Articles/TIS_CtRoomTrial2Art.HTM, last accessed 06/03/07.

Lee, J. (1984) The Law Fact Distinction – From Trial by Ordeal to trial by jury' 12 *AIPLA Q J* p.288.

Leeds Mercury (1854) 'The Leeds Town Hall', Saturday July 22, Issue 6294.

Leeds Mercury (1858) 'Leeds Town Hall Organ', Saturday August 28, Issue 6839.

Lemmings, D. (2000) *Professors of the Law, Barristers and English legal culture in the eighteenth century*, Oxford: Oxford University Press.

Lemmings, D. (2003) 'Ritual, Majesty and Mystery: Collective life and culture among English Barristers, Serjeants and Judges c1500–c1830', in *Lawyers and Vampires*, edited by Pue, W. and Sugarman, D. Oxford: Hart Publishing.

Le Roux, W. (2004) 'Bridges, Clearings and Labyrinths: the architectural framing of post apartheid constitutionalism', 19 *SAPR*, pp.629–45.

Levy, H. (1967) *The Press Council, History, Procedures and Cases*, Macmillan, London.

Lewis, J. (1982) *The Victorian Bar*, London: Robert Hale.

Limbach, J. (2004) 'Working at the Federal Constitutional Court' in Burklin, T., Limbach, J., and Wilkens, M., *The Federal Constitutional Court of Germany: Architecture and Jurisdiction*, Basel: Birkhauser.

Linstrum, D. (1999) *Towers and Colonnades: The architecture of Cuthbert Broderick*, Leeds: The Leeds Philosophical and Literary Society.

Lloyd's Weekly Newspaper, 'Magisterial Complaint of the Police', 18 April 1869, Issue 1378.

Loeffler, J., (1999) 'New designs of embassies and courthouses expose the politics of architecture. But are architects political enough?', Architectural Record, pp.33–5.

Lord Chancellor's Department (1990) *Design Guide and Reference Cost System*, London: Lord Chancellor's Department.

Lord Chancellor's Department (1993) *Court Standards and Design Guide*, London: Lord Chancellor's Department.

Lund, T. (1960) *A Guide to the Professional Conduct and Etiquette of Solicitors*, London: The Law Society.

Macdonald, K. (1995) *The sociology of the professions*, London: Sage.

MacKinnon, F. (1940) *On Circuit 1924–1937*, Cambridge: Cambridge University Press.

194   Bibliography

Maitland, F. (1883–84) 'From the old law courts to the new', *English Magazine*, Vol. 1, pp.3–15. The article is reproduced in *The Cambridge Law Journal*, Vol. 8, no. 1 (1942), pp.2–14.

Maitland, F. (1911) 'Collected papers of Frederick William Maitland' in Fischer, H. (ed), *Law and Jurisprudence*, Cambridge: Cambridge University Press.

Manchester Guardian (1853) 9 April.

Manchester Guardian (1863) 14 September p.3.

Manchester Guardian (1864a) 7 June.

Manchester Guardian (1864b) 26 July p.4.

Manchester Times (1859) 'Manchester Assize Courts – Selection of a Design', 23 April, p.4, cols 1–2.

Manchester Times (1864a) 'A lawyer's opinion of the Manchester Assize Courts', Saturday 6, p.7.

Manchester Times (1864b) Saturday July 30, p.3.

Mason, T. (1989) 'Alice Walker's The Third Life of Grange Copeland: The Dynamics of Enclosure', *Callaloo*, No. 39 (Spring), pp.297–309.

Massey, D. (2005) *For Space*, London: Sage Publications.

Matthews, B. (1927) 'The Woman Juror', *Women Lawyers' Journal*, Vol. XV, No. 2. See http://womenslegalhistory.stanford.edu/articles/juror.htm

May, A. (2006) *The Bar and the Old Bailey 1750–1850*, Chapel Hill: University of North Carolina Press.

McBarnet, D. 'Victim in the Witness Box: Confronting Victimology's Stereotype', (1983) 7 *Contemporary Crises* 293–303.

McEwan, J. (1998) 'The Evidence of Children and Other Vulnerable Witnesses: a Discussion Paper', *International Journal of Evidence and Proof* 32.

McEwan, J. (2000) 'In Defence of Vulnerable Witnesses: the Youth Justice and Criminal Evidence Act 1999', *International Journal of Evidence and Proof* 1–30.

McNally, P. (2009) 'Director of public prosecutions backs cameras in court', *Press Gazette*, 12 January, http://www.pressgazette.co.uk/story.asp?storycode=42781.

McNamara, M. (2004) *From Tavern to Courthouse: Architecture and Ritual in American Law 1658–1860*, Baltimore: Johns Hopkins University Press.

McNeill, D. (2009) 'New Songdo City: Atlantis of the Far East', *The Independent*, 22 June. http://www.independent.co.uk/news/world/asia/new-songdo-city-atlantis-of-the-far-east-1712252.html

McQueen, R. (2003) 'Together We Fall, Divided We Stand: The Victorian Legal Profession in Crisis 1890–1940' in *Lawyers and Vampires*, edited by Pue, W. and Sugarman, D. Oxford: Hart Publishing.

Melhuish, C. (1996) 'Ada Melamede and Ram Karmi: Supreme Court of Jerusalem and House in Tel', *Architectural Design*, 66 nos 11–12, pp.35–9.

Miller, F. (1995) 'The Adversarial Myth', *New Law Journal*, 145, No. 6696 p.734.

Ministry of Justice (2007), *Confidence and confidentiality: openness in family courts, a new approach*, cm 7131, London: Ministry of Justice.

Mitchell, W. (1996) *City of Bits: Space, Place and Infobahn*, Cambridge, MA: MIT Press.

Mohr, R. (1999) 'In between power and procedure: where the court meets the public sphere', *Journal of social change and critical inquiry*, November, number one. http://www.uow.edu.au/arts/joscci/index.html.

Morning Chronicle (1839) 'Assize Intelligence, Northern Circuit Newcastle', 31 July, Issue 21742.

Bibliography | 95

Morning Chronicle (1840) 30 June, Issue 22027.

Mulcahy, L. (2005) 'Feminist Fever? Cultures of Adversarialism in the aftermath of the Woolf Reforms', *Current Legal Problems*, Oxford: Oxford University Press, Vol. 58 edited by Jane Holder and Colm O'Cinneide, pp.215–34.

Mulcahy, L. (2007) 'Architects of Justice: the politics of court house design', *Social and Legal Studies*, Vol. 16(3), pp.383–403.

Mulcahy, L. (2008a) 'The unbearable lightness of being – Shifts towards the virtual trial', *Journal of Law and Society*, Vol. 35(4), pp.464–89.

Mulcahy, L. (2008b) 'Architectural Precedent: Manchester Assize Courts and Monuments to law in the mid Victorian era', 19 *Kings Law Journal*, pp.525–50.

Mulholland, M. (2003) 'Trials in manorial courts in late medieval England' in *Judicial Tribunals in England and Europe 1200–1700, The trial in history, vol. 1* edited by Mulholland, M. and Pullan, B. with Pullan A. Manchester: Manchester University Press.

Murray, G. (1972) *The Press and the Public: The Story of the British Press Council*, London and Amsterdam: South Illinois Press.

Nead, L. (2002) 'Visual culture of the courtroom – reflections on History, Law and the Image, *Visual Culture in Britain*, Vol. 3(2), pp.119–41.

Nicolson, D. and Webb, J. (2000) *Professional Legal Ethics: Critical Interrogations*, Oxford: Oxford University Press.

Nield, B. (1972) *Farewell to the Assizes, The Sixty One Towns*, London: Garnstone Press.

Nugent, Lord (1848) *House of Commons Debate*, February, Vol. 96, cc 368–84.

Office for Criminal Justice Reform (2006) *Convicting Rapists and Protecting Victims – Justice for Victims of Rape: A Consultation Paper*, London: Office for Criminal Justice Reform.

Old Bailey (2009) http://www.oldbaileyonline.org/static/The-old-bailey.jsp Last accessed January 2009.

Oldham, J. (2002) 'Jury research in the English reports in CD-Rom', in Cairn, J., and McLeod, G., *The Dearest Birth Right of the People of England, The jury in the history of the common law*, Oxford: Hart Publishing.

Oldham, J. (2005) *The varied life of the self informing jury*, London: Selden Society.

Omerod, D. (2007) 'Evidence: Witness Anonymity', *Criminal Law Review*, 71–4.

Packard, F. (1839) *A vindication of the separate system of prison discipline from the misrepresentations of the North American Review*, Philadelphia: Dobson.

Pawley, M. (1998) *Terminal Architecture*, London: Reaktion Books Ltd.

Penny Paper and Illustrated Times (1891) 'Birmingham's New Palace of Justice', Saturday July 25, p.960, Issue 1573.

Pevsner, N. (1976), *A History of Building Types*, London: Thames and Hudson.

Philips, T. and Griebel, M. (2003) *Justice Facilities*, Hoboken, NJ: John Wiley and Sons.

Picard, L. (2003) *Elizabeth's London: Everyday life in Elizabethan London*, London: Phoenix.

Pinfold, C. (1963) 'Brentford County Court', *The Builder*, March 15, pp.533–5.

Platt, C. (1995) *Medieval England – A social history and archaeology from the conquest to 1600AD*, Bel Air: Scribner Book Company.

Plotnikoff, J. and Woolfson, R. (2008) 'Making the Best Use of Intermediary Special Measures at Trial', *Criminal Law Review* (2) 91–104.

Pogorzelski, W. and Brewer, T. (2009) 'Cameras in Court: An Empirical Analysis of Television News Coverage of the Trial of Four Police Officers Charged in Amadou Diallo's Death', Paper presented at the annual meeting of the The Law and Society, J.W. Marriott Resort, Las Vegas, NV.

196 Bibliography

Polden, P. (1997) 'Judicial Selkirks: the county court judges and the press 1847–80' in *Communities and Courts in Britain 1150–1900*, (eds) Brooks, C. and Lobban, M., London: Hambledon Press.

Pole, J. (2002) 'Representation and moral agency in the Anglo-American Jury' in *The Dearest birthright of the people of England: The jury in the history of the common law*, (ed) Cairns, J. and McLeod, G., Oxford: Hart Publishing.

Port, M. (1968) 'The new law courts competition 1866–67', *Architectural History*, Vol. 11, pp.75–120.

Prest, W. (1986) *Rise of the barristers, A social history of the English Bar 1590–1640*, Oxford: Clarendon.

Prest, W. (1987) 'Lawyers' in *The Professions in Early Modern England*, edited by Wilfred Prest, Beckenham: Croom Helm.

Preston Guardian (1864) Saturday 30 July, p.4 col 4.

Public Building Service (1997a) *Standard level features and finishes for US court facilities*, Washington, DC: US General Services Administration.

Public Building Service (1997b) *Green Courthouse*, Washington, DC: US General Services Administration.

Quinlan, S. (2009) 'Manchester Civil Justice Centre', Architecture and Law Conference, University of Lincoln, November 2009.

Radul, J. (2007) 'What was Behind Me Now Faces Me – Performance, Staging, and Technology in the Court of Law', http://www.eurozine.com/articles/2007-05-02-radul-en.html, last accessed 28/07/08.

Raffield, P. (2004) *Images and Cultures of Law in Early Modern England – Justice and Political Power, 1558–1660*, Cambridge: Cambridge University Press.

Rendall, J. (1999) 'Gender' in *Gender Space Architecture: An Interdisciplinary Introduction*, London: Routledge.

Resnik, J. and Curtis, D. (2007) 'From "Rites" to "Rights" of Audience: The Utilities and Contingencies of the Public's Role in Court Based Processes' in Masson, A., and O'Connor, K. (eds), *Representation of Justice*, Bruxelles: Peter Lang.

Resnik, J. and Curtis, D. (2010) *Representing Justice: The Creation and Fragility of Courts in Democracies*, New Haven: Yale University Press.

Reynold's Newspaper (1838) On the Housetop or hidden – which? 2 October, p4.

Riddell, R. (2004) 'Robert Smirke' in *Oxford Dictionary of National Biography*, Vol. 51 edited by H. Matthew and B. Harrison, Oxford University Press, Oxford.

Roberts, P., Cooper, D. and Judge, S. (2005) 'Monitoring Success, Accounting for Failure: The Outcome of Prosecutors' Applications for Special Measures Directions under the Youth Justice and Criminal Evidence Act 1999', 9(4) *International Journal of Evidence and Proof* 269.

Roberts, P. and Zuckerman, A. (2004) *Criminal Evidence*, Oxford: Oxford University Press.

Rock, P. (1991) 'Witnesses and Space in a Crown Court', *British Journal of Criminology*, Vol. 31(3), pp.266–79.

Rock, P. (1993) *The Social World of an English Crown Court*, Oxford: Clarendon Press.

Roggee, O. (1968–9) 'Book review: Origins of the Fifth amendment by Leonard W. Levy' 67 *Mich L.Rev* p.862.

Rosen, L. (1966) 'Should the dock be abolished?' 29 *Modern Law Review*, p.289.

Rowden, E. (2009) 'Virtual courts or summary justice – merely brief or unjust?' Architecture and Law Conference, University of Lincoln, November 2009.

Royal Commission on Buildings and Plans for new Courts of Justice (1871) Command 290, London: George Edward Eyre and William Spottiswoods for HMSO. See also House of Commons Parliamentary papers online.

Royal Commission on Assizes and Quarter Sessions (1971) Written evidence submitted to the Commission under the chairmanship of Lord Beeching, Royal Commission on Assizes and Quarter Sessions, 1966–69, London: HMSO.

Royal Commission on Legal Services (1979) Cmnd 7648, London: HMSO.

Royal Courts of Justice (2005) Guide to the Royal Courts of Justice, London: Royal Courts of Justice.

Rubin, G. (1984a) 'Law, Poverty and Imprisonment for Debt, 1869–1914', pp.241–299 in Rubin, G.R., and Sugarman, D. *Law, Economy and Society 1750–1914: Essays in the History of English Law*, London: Professional Books Ltd.

Rubin, G. (1984b) 'The County courts and the tally trade 1846–1914', pp.321–48 in Rubin, G. and Sugarman, D., *Law, Economy and Society 1750–1914: Essays in the History of English Law*, London: Professional Books Ltd.

Rumney, P. (2001) 'Male Rape in the Courtroom: Issues and Concerns', *Criminal Law Review* 205–13.

Sage, H. (1995) 'Do You Receive Me?' *The Lawyer*, 12 September, p.19.

Sanford, B. (1999) 'No contest' in Giles, R. and Snyder, R. (eds), *Covering the Courts: Free Press, Fair Trials and Journalistic Performance*, New Brunswick: Transaction Publishers.

SAVE Britain's Heritage (2004) *Silence in court: The future of the UK's historic law courts*, London: SAVE Britain's Heritage.

Schramm, J. (2000) *Testimony and Advocacy in Victorian Law, Literature and Theology*, Cambridge: Cambridge University Press.

Scott, W. (1831) 'Introduction', in *Anne of Gieierstein or The Maiden of the Mist*, eBooks@ Adelaide http://ebooks.adelaide.edu.au/scott/walter/anne/index.html (Last accessed January 2010).

Scottish Government (1995) 'Live Television Link: an Evaluation of its Use by Child Witnesses in Scottish Criminal Trials', *Crime and Criminal Justice Research Findings No. 4.*

Scully, V. (1993) *The Earth, the Temple and the Gods: Greek Sacred Architecture*, New Haven: Yale University Press.

Sennett, R. (1974) *The Fall of Public Man*, London: Norton and Co.

Sharon, J. (1993) *The Supreme Court Building, Jerusalem* (trans. A Mahler), Jerusalem: Joel Bidan Ltd.

Sharples, J. (2004) *Pevsner Architectural Guides: Liverpool*, New Haven and London: Yale University Press.

Shepard, S. (2006) 'Should the criminal defendant be assigned a seat in court?' *Yale Law Journal*, 115: 2203–10.

Smart, L. and Smart, B. (2009) 'Paramatta Justice Precinct', *Architecture Australia*, Sept/ Oct, pp.77–85.

Smirke, E. (1867) 'The late Sir Robert Smirke, Architect', *The Builder*, August 17, pp.604–6.

Smirke, R. (1824) *Particulars and estimate of the designs for the new courthouse at Maidstone for the county of Kent*, Reference Q/GAc2, Centre for Kentish Studies.

Sobel, W. (ed) (1993) *The American Courthouse: Planning and Design for the Judicial Process*, Chicago: American Bar Association.

198 Bibliography

Stansfield, K. (1985), 'Holding the tiger by the tail', *Construction*, 69, pp.15–17.

Stein, P. (1984) *Legal Institutions, the development of dispute settlement*, London: Butterworths.

Sterling, H. (1963) Plymouth City Law Courts, *The Builder*, October 4, pp.667–670.

Stevens, C. (2002) *Georgian Architecture*, Newton Abbott: David and Charles.

Stone, M. (1991) 'Instant Lie Detection? Demeanour and Credibility in Criminal Trials', *Criminal Law Review* 821.

Stretton, T. (1998), *Women Waging Law in Elizabethan England*, Cambridge: Cambridge University Press.

Studnicki, S. and Apol, J. (2002) 'Witness detention and intimidation: The history and future of material witness law', 76 *St. John's Law Review*, p.483.

Summerson, J. (1970) *Victorian Architecture, Four Studies in Evaluation*, New York: Columbia University Press.

Summerson, J. (1986a) *The Architecture of the eighteenth century*, London: Thames and Hudson.

Summerson, J. (1986b) *Architecture in Britain 1530–1830* (9th edition), London and New Haven: Yale University Press.

Susoeff, S. (1984–85) 'Assessing Children's Best Interests when a Parent is Gay or Lesbian: Toward a Rational Custody Standard', 32 *UCLA L. Rev.* 852.

Tague, P. (2007) 'Barristers' selfish incentives in counselling defendants over the choice of plea', *Criminal Law Review* 3–22.

Tait, D. (2009) 'Democratic spaces in a citadel of authority', *Architecture Australia*, Vol. 98(5), pp.45–6.

Taylor, N., and Joudo, J. (2005) *The impact of pre-recorded video and closed circuit television testimony by adult sexual assault complainants on jury decision-making: an experimental study*. Research and public policy series no. 68. Canberra: Australian Institute of Criminology http://www.aic.gov.au/publications/rpp/68/

Temkin, J. (2000) 'Prosecuting and Defending Rape: Perspectives from the Bar', 25(2) *Journal of Law and Society* 219–48.

Thompson, P. (1967) 'Building of the Year: The New Law Courts, London, 1867–82', *Victorian Studies*, 11, pp.83–6.

Tigar, M. and Levy, M. (1977) *Law and the Rise of Capitalism*, New York: Monthly Review Press.

Times, The (1848) 'Court of Bankruptcy' 15 May.

Times, The (1905) Wednesday 23 June, p.5, col a.

Tittler, R. (1991) *Architecture and power: the town hall and the English urban community c1500–1640*, Oxford: Clarendon Press.

Tschumi, B. (1996) *Architecture and Disjunction*, Cambridge, MA: The MIT Press.

Underwood, R. (1993) 'False witness: A lawyers history of the Law of Perjury', 10 *Arizona Journal of International and Comparative Law*, 215–52.

Wade, A., Lawson, J. and Aldridge, P. (1998) 'Research Stories in Court – Video-taped Interviews and the Production of Children's Testimony', *Child and Family Law Quarterly* 178.

Ware, D. (1949), 'Official Architects of the past: Sir Robert Smirke', *Official Architect*, Vol. 12, pp.682–3.

Waterhouse, A. (1864–5) 'A description of the Manchester Assize Courts', *RIBA papers*, pp.165–76.

Watkins, D. (1982) *The Buildings of Britain: Regency – a guide and gazetteer*, London: Barrie and Jenkins Ltd.

Watson, A. (2001) 'Composing Avebury', *World Archaeology*, Vol. 33, No. 2, *Archaeology and Aesthetics* (Oct), pp.296–314, Published by: Taylor & Francis Ltd.

Weatherhead, P. (1991), 'No stone unturned – refurbishment Maidstone County hall', *Building*, 21 June 1991, pp.42–5.

Western Mail, The (1886) 'London letter', 13 February, Issue 5227.

West-Knights, L. (2005) 'The Courtroom of the Present' http://www.lawonline.cc/.htm, last accessed April 2005, pp.8–9.

Widdison, R. (1997) 'Beyond Woolf: The Virtual Court House', 2 *Web J CLI*.

Williams, R. (1978) 'The Press and popular culture' in *Newspaper History from the 17th century to the present day*, (eds) Boyce, G., Curran, J. and Wingate, P., London: Constable Press.

Wong, F. (ed) (2001) *Judicial administration and space management: A guide for architects, court administrators and planners*, Gainesville: University Press of Florida.

Wright, G. (2004) 'Due Process', *Building Design and Construction*, Vol. 45, Issue 5, May, pp.34–43.

Wright, M. (1996) *Justice for victims and offenders: A restorative response to crime* (2nd edition), Winchester: Waterside Press.

Index

Accused (see also defendant) 18, 45, 50, 51, 52–3, 59–79, 151–2, 165, 167, 172, 174–7
Achilles shield 15, 84
Act of settlement 5, 127
administrative function 25–6
adversarial 11, 12, 46, 52, 62–4, 67, 80n13, 85, 94, 100, 162, 163, 165, 173, 176–7
Alexander, Daniel 114
Antwerp law courts 148, 152, 153
Ardlamont divorce case 100
Arches 29–31
Assizes 5–6, 24, 25–6, 27, 28, 32, 33, 43, 54, 63, 74, 84–5, 88, 114, 116–17, 119, 126, 128, 140–1, 147
Assize courts: Bristol 117; Carlisle 33, 48; Devizes 117; Durham 45, 101; Liverpool 31, 48, 70, 120, 127, 129 (see also St Georges Hall); Manchester 48, 54, 55, 62, 66, 89, 90, 118, 122–4, 126, 127, 130, 132, 169; Reading 116, 117; York 5, 31, 33, 47, 48, 55, 68, 89, 113
Attorney General 60, 66, 95
Attorneys 26, 43,46, 49, 51, 54, 60–2, 65, 66, 67, 78, 87 (see also lawyers)
Authority 1, 7, 9, 21, 28, 89, 93, 131, 134, 145–6,151, 169, 172

back stage 56, 57n3
bar (physical) 9, 38, 39–41, 45, 46, 48, 59, 60–1, 65, 68, 71, 73–4, 77, 78, 87, 151
Barristers 9, 23–4, 46, 48, 49, 52, 54, 59, 60–7, 79n4, 81n16, 88, 90, 102, 105, 108n13, 124 (see also Counsel)
Bar library 48, 49, 62, 66, 67
Barry, Charles 115, 118, 123
battle of the styles 117, 136n20

Beeching Commission 50, 140–1
Bell, Ingress 123
Bentham, Jeremy 71, 81n23, 85, 87, 89, 94
Boothall, Gloucestershire 27
Borough courts 128
Bordeaux Law Courts 148, 154
Brisbane supreme court 20, 155
Broderick, Cuthbert 121
Brougham, Henry 121, 128
Brussels palais de justice 152
Butler, John Dixon 117

Cameras in court 97, 104–6, 162–78
Capitalism 54, 131, 133, 176
Carr, John 5, 33, 48, 89, 115, 116
Castles 3, 24, 26, 112
centralization of design 140–5
ceremony 17–21, 27–29 (see also ritual)
Chancery 39, 114, 128, 129
Chartism 92, 131
children in court 56, 70, 94, 165
Churches 3, 25, 47, 115, 131
circuits 25, 26, 28, 61–2
Circulation routes 1, 9, 20, 24, 27, 33, 48–51, 52, 54–7, 89, 90, 92, 149, 153, 159, 160 (see also segregation)
City of Justice, Barcelona 148
Civic pride 8, 112, 124–6, 127, 131, 134
civil procedure rules 170
Civil liberties 1, 2, 5, 38
Clerk 10, 41, 43, 45, 46, 48, 49, 51, 61, 62, 64, 67, 75, 88, 93, 113, 128
coat of arms 41, 140, 170
Comfort 8, 27, 56, 65, 67, 89, 112, 113–18, 126, 142
common law 7, 11, 19, 21, 62, 85, 98, 118, 128, 129, 129, 133, 139, 160, 165, 167, 167, 175

202 Index

common pleas 18, 39, 46–7, 60, 114
Commonwealth courts Melbourne 155, 156–7, 158
community justice centres 6, 18, 32, 139, 151, 157
concentric ring 48, 49, 50, 90
(*see also* circulation, segregation)
consistory courts 26, 41, 45
constitutional court South Africa 17, 139, 154, 155, 156
consultation rooms 48, 49, 67, 123, 145
containment 1, 9, 10, 21–22, 38, 55, 60, 93, 107, 150
contamination 9, 83
corporations 25, 43
counsel 2, 9, 11, 21, 46, 50, 52, 53, 55, 59–82, 87, 88, 90, 105, 159, 163
council houses 24
county halls 24, 25, 31, 47–8; Abingdon 29, 30; Kingston 31; Lincoln 90; Presteigne 43, 44
county courts 6, 17, 25, 92–4, 102, 116, 117, 129, 133, 134, 141, 146, 156
court standards and design guide 14, 23, 46, 102, 169, 170, 174
court of appeal 166, 167
court of exchequer 39, 74, 78
courts of equity 175
court of star chamber 25, 85
court of wards and liveries 41
cross examination 12, 63, 163, 167, 175
crown court 10, 27, 31, 41, 46, 47, 48, 60, 71, 76, 78, 96, 141, 142, 166, 175; Aylesbury 173; Birmingham 96; Coventry 96; Dorchester 45, 101; Lincoln 147; Southwark Crown court 72; Snaresbrook 97; Wood Green 23, 53, 95, 102
crown pleas 19
current state of building stock 145–51

dais 17, 19, 39, 41, 55, 64, 144, 169
defendant 1, 2, 6, 9, 10, 11, 18, 21, 33, 38, 39, 41, 45, 46, 48, 52, 59–82, 84, 86, 89, 96, 97, 100, 103, 104, 106, 151–2, 155, 160, 165, 166, 167, 168, 169, 172, 173, 174–5 (*see also* accused)
Degradation 9, 74–8, 106, 131–2
Democratic 12, 71, 89, 95, 96, 99, 106, 107, 131, 141–2, 146, 151–9, 160, 163
Differentiation 39, 43, 55, 60–4, 71
Dignity 1, 9, 10, 12, 67, 75, 78, 87, 93, 106, 114, 143, 151, 152, 173

Dining rooms 10, 48, 54, 119, 123
Dock 1, 2, 9, 10, 32, 33, 38, 59, 60, 68–73, 74–9, 82n30, n42, 96, 103, 131, 151–2, 172
doctors commons 87
Due process 9, 12, 21, 38, 42, 51–2, 59, 77, 149, 160, 168

Ecclesiatical 19, 26, 128, 130
Elmes, Harvey Lonsdale 5, 120–1
Entrance hall 23, 33, 55, 66, 90, 92, 96, 148, 150, 157, 173
Evidence, law of 5, 9, 51–3, 63, 68, 128, 164–7, 172, 175
European Court of Human Rights 148, 153, 167
European Court of Justice 153

fear of unrest 9, 92, 131–4
Federal Constitutional courts Germany 3, 152
female spectators 95, 104
Free field court of corbey 28, 39
French revolution 131
front stage 49, 56

Glass 10, 14, 20, 50, 68, 76, 78, 79, 96, 151, 152–3, 159, 177
gothic 5, 14, 114, 117–18, 119, 134, 142, 147, 160
Greek courts 22, 84
Guilds 25, 26
Guildhalls 24, 26

Halls 18–19, 21, 24–5, 26, 27, 29–32, 33, 35n14, 39, 41, 43 (*see also* county halls, shirehalls, town halls, markethalls, moothalls, Westminister hall, Winchester hall)
Harrison, Thomas 48, 116, 117
High Courts of Justice, Madrid 148
Houses of Parliament 117, 119, 127
Howard league for penal reform 32, 68, 77
Human rights 56, 75, 85, 167
hundred court 17, 27

in camera 94, 99
industrialisation 54, 112, 115, 124–6, 130, 131–2, 142
inns of court 22, 47, 61–2
inquisitorial 62, 85
insiders1, 49, 102

Index   203

International Criminal Tribunal for the former Yugoslavia 172
Intimidation 9, 53, 83, 96, 151, 168
Isolation 10, 64, 67–74, 79, 178
Israeli supreme court 3, 23

Judge 5, 6, 7, 10, 11, 20, 25–6, 27, 28–9, 33, 39, 41, 45, 46, 47, 48, 49, 50, 52, 53, 54, 60, 63, 65, 67, 71, 76, 78, 83, 84, 85, 86, 88, 89, 90, 93, 94, 95, 96, 97, 98, 99, 100, 103, 104, 105, 114, 116, 121, 123, 124, 130, 134, 141, 145, 151, 153, 157, 160, 163, 164, 166, 168, 170, 175
Jury 19, 25, 26, 27, 33, 39, 41, 43, 45–6, 49, 50, 51, 54, 55, 61, 63, 66, 67, 74, 75, 77, 78, 83, 87, 95, 96, 100, 101, 102, 103, 107, 108n12, 129, 144, 147, 151, 165, 169, 175
justices of the peace 25, 26, 84, 129, 132

Kalgoorlie Courts Project 154
Katseiris, Paul 143, 155, 178
Kings bench, court of, 24
Kohn, Bernard 153, 154, 155–6

Law Society 75, 77, 80n14
Lawyers 3, 4, 9, 39, 41, 46, 48, 49, 52, 54, 55, 59–67, 73, 77, 78, 90, 92, 93, 100, 102, 104, 114, 118, 119, 122, 132, 133, 163, 164, 174, 175, 177
Le Corbusier 143
Legitimacy 1, 9, 10, 22, 23, 84, 86, 98, 99, 146, 175
Light 17, 19, 20, 124, 144, 145, 154, 155, 156
Linear circulation 6, 49, 50 (*see also* circulation, segmentations, zones)
Lloyd Wright, Frank 32

Magistrates court 95, 102, 141, 143, 145, 147, 148, 154, 166, 169, 173
magna carta 18, 26, 84
Manchester Assize courts 48, 49, 54, 55, 62, 66, 89, 90, 101, 118, 122–4, 125, 126, 127, 129, 130, 132, 169
Manchester civil justice centre 14, 50, 145, 148, 156, 157, 159, 171
manorial courts 17, 25, 26, 43, 84, 128
Market cross 20, 26, 35n14
Market place/halls 7, 20, 24, 25, 26–7, 112, 114, 133
Methodology 5–6
Modernism 142–4

Moothills 6, 15
Multi purpose halls 10, 26, 27, 31, 32, 34, 43

Neo-classical 50, 117, 118, 142
nisi pruis 27, 32, 74

Old Bailey 3, 20, 33, 45, 48, 52, 69, 74, 89, 90, 173
Open court 10, 11, 20, 31, 84–97, 104
open space 20, 22, 27, 155
open trial 86, 97, 99
orality 94, 164, 166, 175

pageantry 28, 29
panopticon 71
parallel functions 29–32, 50, 153
Parramatta Justice Precinct 155
Participatory process 1, 2, 7, 12, 18, 38, 59, 74, 83, 86, 87, 93–4, 96–97, 103, 106, 139, 152, 162, 163, 167, 171, 173
Petty sessions 4, 123, 128
Photography 3, 150 (*see also* cameras)
police station 31, 72, 166, 172
Press 2, 10, 11, 23, 45, 49, 83–4, 85, 90, 97–111, 114, 123
Presumption of innocence 2, 10, 60, 72, 75, 76, 77, 78
professional etiquette 9, 53, 61–2
Private finance initiative 147
Publicity 23, 38, 85, 87–95, 105
Public Gallery 6, 49, 88, 89, 90, 97, 104, 106, 152, 163, 173 (*see also* spectators)
Public Hearing 75, 84–95, 97–101, 167, 170
Public interest 10, 11, 45, 84, 97, 98, 99, 101, 103, 132
Public space/sphere 1, 2, 3, 5, 6, 7, 8, 9, 12–13, 22–3, 24–7; 31–2, 38–9, 43, 49–50, 54, 55–6, 59–60, 83–4, 95–7, 103–7, 112, 118, 119, 130–1, 139–40, 151–61, 175–8
purpose built courts 33

quarter sessions 4, 25, 32, 116, 123, 140, 141

reform of legal system 8, 9, 12, 52, 63, 67–8, 112–13, 121, 126, 127–9, 134–5, 140–2, 153, 163, 165
Rehabilitation 18, 67, 151
restorative Justice 6
Ritual 1, 2, 7, 9, 10, 11, 12, 18, 19, 21, 27–9, 53, 71, 85, 86, 93, 103, 162, 163, 169, 173, 173, 174, 178

204 Index

Robing rooms 10, 54, 66, 123, 124
Rogers, Richard 148, 152, 153, 154
Royal courts of justice, London 3, 5, 14, 19, 29, 49, 56, 66, 92, 104, 114, 118, 119, 127, 129, 131, 145

Sacred 15, 22, 39–43, 84
St Georges hall Liverpool cover 5, 6, 31, 70, 119, 120, 121, 122, 127, 133, 169
Salle de pas perdu 50, 121, 154 (*see also* entrance)
Security 51, 60, 72, 77, 77, 78, 131, 150, 151, 156, 160, 170, 173
segmentation 2, 9, 33, 38–57, 59, 60, 83, 88, 95, 154
segregation (*see also* circulation) 2, 9, 38–57, 59, 60, 83, 86, 106–7, 142, 147, 151, 153, 174
separate system 68
sessions house or court 20, 117; Beverley 33; Birkenhead 117; Liverpool 90, 91, 117; Maidstone 116; Middlesex 33, 113; Northampton 27, 31, 48; Spilsby 117
Sheriff 26, 28, 29, 49, 89
Shire halls 24, 26, 112, 113, 116; Bedford 29, 117; Bodmin 6; Derby 33; Ely 33; Nottingham 6, 45, 89; Warwick 29, 61, 68
Sightlines 10, 96, 107, 170
Silent system 68, 81n23
Smirke, Robert 5, 33, 48, 54, 55, 90, 115, 116, 117, 147
Soane, John 50, 65, 114, 115, 117, 119, 124, 127
Social court 9
Sole use 24, 33, 47, 48, 67, 114, 122
Solicitors 46, 46, 50, 52, 61, 62, 66, 67, 75, 92 (*see also* attorneys and lawyers)
Special measures 53, 165, 168
Spectacle 17, 52, 56
Spectators 1, 10, 27, 28, 39, 41, 45, 51, 52, 55, 66, 77, 84, 84–94, 95–104, 105–7, 132, 175, 177, 178 (*see also* public)
Stone circles 6, 15, 19
Street, George Edmund 5, 14, 49, 92, 115, 119, 127

Surveillance 10, 38, 55, 83, 93, 96, 106, 106–7, 150, 153, 162
symbolism 19, 115, 124–6, 152, 174

Table 18, 39, 41, 43, 45, 46, 60, 61, 62, 64–5, 73, 74, 78, 93, 147
technology 11–12, 21, 53, 72, 154, 156, 162–78
televising trials 11–12, 20, 73, 83, 104–7, 165
teutonic courts 15
theatre 4, 7, 54, 144, 176–7
thresholds 7, 22–4
town halls 3, 24, 26, 27, 43, 47, 112; Birmingham 121; Horsham 116; Leeds 31, 121, 122, 124, 126, 127, 129, 130, 133, 169; Oxford 311; St Albans 69
Trees 6, 7, 16, 17–18, 155
trial by ordeal 19, 22, 53, 87

unrest 9, 92, 131–2
urbanization 115

Victoria law courts, Birmingham 56, 90, 123–4, 126, 127, 131, 169
Video 97, 105, 106, 162–78
virtual courts 7, 53, 72, 97, 101, 162–78

walls 6, 7, 15–17, 19, 20, 21, 29, 34, 52, 87, 89, 101, 114, 144, 145, 150–1, 152, 153, 163, 170
Waterhouse, Alfred 48–9, 54, 66–7, 90, 101, 115, 118, 122–3, 124, 126
Webb, Aston 123
Westminister hall 24, 27, 39, 41, 50, 65, 74, 87, 89, 114, 115, 120, 121
Whaddon illuminations 39, 41–2, 45, 46, 60, 78
Winchester great hall 31, 45, 87, 101
Witnesses (*see also* intimidation)
Wyatt, John 45, 101

Zones 9, 10, 38, 39, 43, 47, 50, 52, 54, 107, 150, 153, 154 (*see also* segregation, circulation)